Table of Contents

About Author 5

Introduction to Bullet Inversion System (BIS) 6

Overview of BIS Concept and Objectives 6

Historical Context and Advancements in Personal Protection Technology 7

Conclusion 8

Fundamental Physics of Electromagnetism 8

Lorentz Force Law and Its Relevance to BIS 9

Maxwell's Equations and Their Application in BIS 9

Conclusion 11

Magnetic Fields and Bullet Deflection 12

Understanding the Behavior of Magnetic Fields 12

Analyzing the Force Exerted on Moving Electric Charges 12

Conclusion 14

Material Science for BIS 15

Properties of Ferromagnetic Materials: Iron, Cobalt, Nickel 15

Challenges Posed by Lead and Non-Magnetic Materials in Bullets 16

Conclusion 17

Electromagnetic Induction 18

Faraday's Law of Induction and its Application 18

Inducing Currents in Bullets to Generate Magnetic Fields 18

Practical Considerations for BIS Implementation 19

Future Research and Development 20

Conclusion 21

Plasma Physics for Bullet Deflection 21

Plasma as an Ionized Gas and Its Properties 21

Creating and Maintaining a High-Energy Plasma Shield 22

Practical Applications and Challenges 23

Future Research and Development 24

Conclusion 24

Power Requirements and Solutions 25

Estimating Power Needs for Generating Strong Magnetic Fields 25

Exploring High-Energy Batteries and Supercapacitors 26

Practical Implementation and Future Directions 27

Conclusion...29

Miniaturization and Portability 29

Challenges in Miniaturizing Magnetic Field Generators and Plasma Systems............29

Designing a Portable BIS Device..30

Future Directions 32

Conclusion...32

Magnetic Field Generators 32

Types and Designs of Electromagnets and Permanent Magnets 33

Utilizing Superconducting Materials for Compact Design.................................34

Future Directions 35

Conclusion...35

Plasma Generation Techniques...36

Conclusion...39

Simulation of Bullet Inversion System (BIS) 39

Conclusion...41

Prototype Development for Bullet Inversion System (BIS)...................................42

Field Testing of Bullet Inversion System (BIS) 44

Safety Testing for the Bullet Inversion System (BIS) 47

Energy Efficiency Improvements for the Bullet Inversion System (BIS) 50

Material Durability and Performance in High-Energy Environments 54

Future Material Research: Innovations in Magnetic Materials and Alternatives for
Bullet Influence 57

Advances in Energy Storage: Innovations in Battery Technology and
Supercapacitors 60

Integration with Protective Technologies: Combining BIS with Advanced Body Armor
and Automated Threat Detection Systems 64

Legal and Ethical Considerations: Addressing BIS Use and Ethical Implications of
Personal Defense Technology 67

Cost Analysis and Feasibility of the Bullet Inversion System (BIS) 71

Impact on Law Enforcement and Military: Potential Applications and Tactical
Operations 74

User Training and Operation for Bullet Inversion System (BIS): Developing Programs and Ensuring Effective Use 78

Public Acceptance and Perception of Bullet Inversion Systems (BIS): Gauging Opinion and Building Trust 81

Environmental Impact and Sustainability of Bullet Inversion Systems (BIS): Analyzing Footprints and Exploring Sustainable Solutions 85

Comparative Analysis with Existing Technologies: Evaluating Bullet Inversion Systems (BIS) Against Current Body Armor and Protective Gear 89

Potential for Non-Lethal Applications of Bullet Inversion Systems (BIS) 93

Case Studies and Historical Precedents in Protective Technologies 96

Technological Challenges and Overcoming Them in Bullet Inversion Systems (BIS) 99

Collaborative Research Opportunities in Bullet Inversion Systems (BIS) 102

Government and Military Funding for Bullet Inversion Systems (BIS) 105

Ethical and Societal Impact of Bullet Inversion Systems (BIS) 109

Intellectual Property and Patents for Bullet Inversion Systems (BIS) 112

Marketing and Commercialization Strategies for Bullet Inversion Systems (BIS) 116

Feedback and Iterative Design for Bullet Inversion Systems (BIS) 119

Workshops and Conferences: Sharing Research and Networking in the Advancement of Technologies 122

Educational Outreach and Awareness: Advancing Knowledge and Engagement on BIS Technology 125

Technological Readiness Levels (TRL): Evaluating and Advancing the Bullet Inversion System (BIS) 129

Cross-Disciplinary Collaboration: Bridging Physics, Engineering, and Materials Science 132

Long-Term Vision and Goals: Shaping the Future of BIS Technology 136

Regulatory Compliance for Bullet Inversion System (BIS) 139

User Interface and Experience Design for Bullet Inversion System (BIS) 142

Robustness and Reliability Testing for Bullet Inversion System (BIS) 145

Cultural and Geopolitical Considerations for Bullet Inversion System (BIS) 148

Scalability and Mass Production of Bullet Inversion Systems (BIS) 151

Continuous Innovation and Research: Fostering a Culture of Progress 154

Funding Models and Investment: Navigating Financial Strategies for BIS Development 158

Disaster Response and Humanitarian Applications: Harnessing BIS Technology for Crisis Management and Civilian Protection 161

Public Policy and Advocacy: Shaping the Future of Bullet Inversion Systems (BIS) 164

Global Collaboration and Standards: Advancing Bullet Inversion Systems (BIS) through International Efforts 168

About Author

Jamil E. Brown is a seasoned IT professional with over seven years of diverse experience in the tech industry. His career spans roles as a Data Center Technician and a Jr. Project Manager, where he has honed his skills in hardware management, network troubleshooting, and server diagnostics. Jamil has worked with renowned companies like Google, Oracle, and Amazon, focusing on server build projects, cloud infrastructure maintenance, and advanced network configurations. He holds an Associate of Science in Information Technology from Northern Virginia Community College, graduating Magna Cum Laude.

Jamil was part of GMU's US ARMY ROTC Program 2010-2013.

For more about Jamil's work and credentials, visit his LinkedIn profile:
https://www.linkedin.com/in/jamil-e-brown/

Preface:

In an era where personal safety and advanced technology increasingly intersect, the concept of a Bullet Inversion System (BIS) presents a groundbreaking leap in protection. This theoretical blueprint aims to explore the feasibility and potential challenges of creating a device capable of redirecting incoming bullets using a reverse magnetic field. The vision for BIS encompasses a comprehensive understanding of physics, electromagnetism, and material science, coupled with cutting-edge advancements in energy storage and miniaturization.

This document serves as a detailed scientific exploration, providing a thorough analysis of the principles and technologies that could make BIS a reality. From the foundational physics of electromagnetism and the complexities of material science to the practical challenges of power requirements and portability, every aspect is meticulously examined. Theoretical frameworks such as electromagnetic induction and plasma physics are investigated for their potential application in influencing bullet trajectories.

Practical considerations, including simulation, prototype development, and safety testing, are outlined to ensure that the BIS not only functions effectively but also maintains user safety. The exploration extends to the broader impacts of BIS technology, addressing legal, ethical, and environmental concerns, as well as the potential for non-lethal applications and integration with other protective systems.

Through collaboration with research institutions, leveraging government and military funding, and navigating the intellectual property landscape, this blueprint outlines a path forward for BIS development. The goal is to inspire continuous innovation and global collaboration, ultimately leading to a reliable, portable device that significantly enhances personal protection beyond the capabilities of current body armor technology.

This preface sets the stage for an in-depth journey into the future of personal safety, where the integration of advanced scientific principles and technological innovations holds the promise of unprecedented protection.

Introduction to Bullet Inversion System (BIS)

In the rapidly evolving landscape of personal protection technology, the Bullet Inversion System (BIS) emerges as a revolutionary concept aimed at significantly enhancing individual safety. The BIS represents an ambitious and innovative approach to personal defense, leveraging advanced principles of electromagnetism, physics, and material science to create a protective shield capable of redirecting incoming bullets. This system aspires to transcend the limitations of traditional body armor by offering a dynamic and proactive defense mechanism that can neutralize threats in real time.

Overview of BIS Concept and Objectives

The core concept of the Bullet Inversion System revolves around the use of a reverse magnetic field to alter the trajectory of bullets, effectively protecting the wearer from harm. Unlike conventional body armor, which relies on the material's ability to absorb and dissipate the kinetic energy of a projectile, the BIS seeks to actively influence the path of the bullet, diverting it away from the intended target. This approach not only reduces the risk of injury but also potentially neutralizes the threat before impact.

The primary objectives of the BIS are multifaceted:

1. **Enhanced Safety**: Provide superior protection against ballistic threats by deflecting or redirecting bullets.
2. **Portability**: Develop a compact and lightweight system that can be easily carried and operated by individuals in various environments.
3. **Energy Efficiency**: Ensure that the system operates efficiently with minimal power consumption, making it practical for extended use.
4. **Scalability**: Design the BIS to be adaptable for different applications, from personal use by civilians to deployment in law enforcement and military contexts.

Historical Context and Advancements in Personal Protection Technology

To appreciate the significance of the BIS, it is essential to understand the historical context of personal protection technology. The quest for effective personal defense mechanisms dates back centuries, with early examples including rudimentary shields and armor made from materials like wood, leather, and metal. Over time, advancements in material science and engineering have led to the development of more sophisticated protective gear.

Early Developments

The evolution of personal armor began with ancient civilizations, where warriors donned leather and metal armor to shield themselves from weapons like swords and arrows. The advent of gunpowder and firearms in the Middle Ages introduced new challenges, necessitating the development of armor capable of withstanding the impact of bullets. This period saw the introduction of plate armor, which offered enhanced protection but at the cost of increased weight and reduced mobility.

Modern Body Armor

The 20th century marked a significant turning point in the development of personal protection technology. The introduction of synthetic materials like Kevlar revolutionized body armor design. Kevlar, a high-strength polymer, provided a lightweight yet robust solution capable of stopping bullets and shrapnel. The adoption of Kevlar in military and law enforcement applications in the 1970s and 1980s significantly improved the survivability of personnel in combat and high-risk situations.

Advancements in Material Science

Recent advancements in material science have further enhanced the capabilities of personal protection gear. Innovations such as carbon nanotubes, graphene, and advanced ceramics have contributed to the development of lighter, stronger, and more flexible armor solutions. These materials offer superior ballistic resistance while minimizing the physical burden on the wearer, thus improving mobility and comfort.

The Role of Electromagnetism and Technology

The integration of electromagnetism and technology into personal protection represents the next frontier in this field. Concepts like active protection systems (APS), which are used in military vehicles to detect and intercept incoming projectiles, demonstrate the potential of combining advanced sensors and electromagnetic fields for defense purposes. The BIS builds on these principles, aiming to miniaturize and adapt them for individual use.

Challenges and Considerations

Despite the promising potential of the BIS, several challenges must be addressed to realize its practical implementation. These include:

1. **Material Compatibility**: Most bullets are made of non-magnetic materials like lead, presenting a challenge for magnetic deflection. Research into alternative methods, such as electromagnetic induction or plasma shielding, is crucial.
2. **Power Requirements**: Generating and maintaining strong magnetic fields or plasma shields requires significant energy. Developing efficient power sources, such as advanced batteries or supercapacitors, is essential.
3. **Portability and Size**: The BIS must be compact and lightweight to be practical for everyday use. Advances in miniaturization and material science will play a key role in achieving this goal.
4. **Safety and Reliability**: Ensuring the safety and reliability of the BIS is paramount. Rigorous testing and validation are necessary to confirm its effectiveness and to prevent unintended consequences.

Conclusion

The Bullet Inversion System represents a bold step forward in personal protection technology, combining theoretical physics, electromagnetism, and cutting-edge material science to create a potentially transformative defense mechanism. By actively redirecting incoming bullets, the BIS aims to provide unparalleled protection, transcending the capabilities of traditional body armor. The historical evolution of personal protection, from ancient armor to modern ballistic gear, underscores the continuous quest for better and more effective defense solutions. The BIS embodies the next stage in this journey, offering a vision of enhanced safety and security through innovative technology. However, significant research and development efforts are needed to overcome the technical challenges and bring this groundbreaking concept to fruition.

Fundamental Physics of Electromagnetism

Electromagnetism, a fundamental branch of physics, describes the interactions between electric charges and magnetic fields. It encompasses a wide array of phenomena that are crucial for understanding and developing advanced technologies, including the Bullet Inversion System (BIS). This system aims to use electromagnetic principles to deflect or redirect bullets, thus providing a new form of personal protection. To fully grasp the theoretical foundation of the BIS, it is essential to delve into the fundamental physics of electromagnetism, focusing on the Lorentz Force Law and Maxwell's Equations.

Lorentz Force Law and Its Relevance to BIS

The Lorentz Force Law is a cornerstone of electromagnetism, describing the force experienced by a charged particle in the presence of electric and magnetic fields. Mathematically, the Lorentz Force $\mathbf{F}$ is given by:

$$\mathbf{F} = q(\mathbf{E} + \mathbf{v} \times \mathbf{B})$$

where q is the electric charge of the particle, $\mathbf{E}$ is the electric field, $\mathbf{v}$ is the velocity of the particle, and $\mathbf{B}$ is the magnetic field. This equation reveals how a charged particle moves under the influence of both electric and magnetic fields, making it directly relevant to the functioning of the BIS.

Relevance to BIS

In the context of the BIS, the Lorentz Force Law helps explain how a magnetic field can be used to alter the trajectory of a bullet. Bullets, typically made of materials like lead, are not inherently charged particles. However, if we can induce a current in the bullet or if the bullet is composed of or coated with a material that interacts with magnetic fields, the Lorentz Force can then influence its path. This involves creating a situation where the bullet's interaction with a magnetic field generates a force that changes its direction, effectively deflecting it away from the target.

Induced Currents and Magnetic Interaction

One approach is to induce an electric current in the bullet as it passes through a magnetic field. According to Faraday's Law of Induction, a changing magnetic field can induce an electromotive force (EMF) in a conductor. This induced EMF creates a current, which in turn interacts with the magnetic field to produce a force on the bullet, according to the Lorentz Force Law. This force can be designed to counteract the bullet's trajectory, redirecting it away from the person being protected.

Maxwell's Equations and Their Application in BIS

Maxwell's Equations are a set of four fundamental equations that describe the behavior of electric and magnetic fields and their interactions with matter. These equations form the theoretical backbone of electromagnetism and are essential for understanding how the BIS could be developed and implemented.

1. **Gauss's Law for Electricity:**

$$\nabla \cdot \mathbf{E} = \frac{\rho}{\epsilon_0}$$

This equation states that the electric flux through a closed surface is proportional to the charge enclosed within that surface. It describes how electric charges generate electric fields.

2. **Gauss's Law for Magnetism**:

$$\nabla \cdot \mathbf{B} = 0$$

This law indicates that there are no magnetic monopoles; magnetic field lines are continuous and have no beginning or end, forming closed loops.

3. **Faraday's Law of Induction**:

$$\partial t \nabla \times \mathbf{E} = -\frac{\partial \mathbf{B}}{\partial t}$$

This equation describes how a changing magnetic field over time induces an electric field. It is fundamental for understanding electromagnetic induction and is crucial for inducing currents in bullets for the BIS.

4. **Ampère's Law (with Maxwell's correction)**:

$$\partial t \nabla \times \mathbf{B} = \mu_0 \mathbf{J} + \mu_0 \epsilon_0 \frac{\partial \mathbf{E}}{\partial t}$$

This law states that magnetic fields can be generated by electric currents and changing electric fields. It encompasses the idea that both currents and time-varying electric fields can produce magnetic fields.

Application in BIS

Maxwell's Equations provide a comprehensive framework for designing and understanding the BIS. Here's how each equation contributes to the concept:

- **Gauss's Law for Electricity**: This helps in understanding the distribution of electric fields around the BIS and how these fields can be manipulated to influence the path of a bullet.

- **Gauss's Law for Magnetism**: Ensures that the design of the BIS takes into account the continuous nature of magnetic fields. This is crucial for creating stable and predictable magnetic fields that can deflect bullets.

- **Faraday's Law of Induction**: Central to inducing currents in bullets. As a bullet moves through a magnetic field generated by the BIS, a changing magnetic field induces an EMF in the bullet, creating a current that interacts with the magnetic field, resulting in a deflective force.

- **Ampère's Law**: Used to design the magnetic field generators within the BIS. Understanding how currents and changing electric fields generate magnetic fields allows for the creation of strong, controlled magnetic fields necessary for deflecting bullets.

Design Implications

Designing the BIS requires a deep understanding of these principles. For instance, the placement and strength of magnetic field generators must be carefully calculated to create a field strong enough to induce the necessary current in a bullet. Additionally, the system must account for the rapid time scales involved, as bullets travel at high velocities, necessitating quick and precise responses from the magnetic field.

Practical Challenges

Several practical challenges arise when applying these theoretical principles to a real-world BIS:

1. **Material Limitations**: Most bullets are made of non-magnetic materials, such as lead. Finding ways to induce currents in these materials or developing new bullet compositions that interact more readily with magnetic fields is essential.
2. **Energy Requirements**: Generating strong, localized magnetic fields and maintaining them for the necessary duration requires significant energy. Advanced power sources and energy-efficient designs are crucial for making the BIS portable and practical.
3. **Rapid Response Time**: Bullets travel at high speeds, often exceeding the speed of sound. The BIS must respond almost instantaneously to detect and deflect these projectiles. This requires sophisticated sensors and control systems capable of rapid operation.
4. **Size and Portability**: The BIS must be compact enough to be carried by individuals in various settings. This poses challenges in miniaturizing the components while maintaining their effectiveness and reliability.

Conclusion

Theoretical principles of electromagnetism, as encapsulated by the Lorentz Force Law and Maxwell's Equations, provide a robust foundation for developing the Bullet Inversion System. Understanding how electric and magnetic fields interact with materials, inducing currents, and generating forces on projectiles are critical steps towards realizing a practical BIS. Overcoming the practical challenges of material compatibility, energy requirements, response time, and portability will require innovative solutions and advances in technology. By integrating these fundamental principles with cutting-edge research and development, the BIS has the potential to revolutionize personal protection, offering a proactive and dynamic defense mechanism against ballistic threats.

Magnetic Fields and Bullet Deflection

The concept of deflecting bullets using magnetic fields is both fascinating and complex, involving a deep understanding of the behavior of magnetic fields and the forces they exert on moving electric charges. This section will explore these topics in detail, providing both theoretical insights and practical considerations for the development of technologies like the Bullet Inversion System (BIS).

Understanding the Behavior of Magnetic Fields

Magnetic fields are a fundamental aspect of electromagnetism, produced by electric currents and changing electric fields. These fields exert forces on moving electric charges and magnetic materials, which can be harnessed for various applications, including the deflection of bullets.

Magnetic Field Generation

Magnetic fields are generated by electric currents flowing through conductors. According to Ampère's Law, a steady current I in a conductor produces a magnetic field B that encircles the conductor. The strength and direction of this field depend on the current's magnitude and the geometry of the conductor.

In the context of the BIS, generating a magnetic field strong enough to deflect a bullet requires significant current and precise control over the field's shape and intensity. Technologies such as superconducting magnets or high-strength electromagnets can be employed to create the necessary magnetic fields.

Magnetic Field Characteristics

Magnetic fields have several key characteristics that influence their interaction with objects:

1. **Field Lines**: Magnetic field lines represent the direction and strength of the field. They form closed loops, emanating from the north pole of a magnet and returning to the south pole. The density of these lines indicates the field's strength.
2. **Field Strength**: The magnetic field strength B (measured in teslas, T) depends on the current generating the field and the distance from the source. Higher currents and closer proximity to the conductor result in stronger fields.
3. **Field Uniformity**: Uniform magnetic fields exert consistent forces over an area, while non-uniform fields can produce varying forces. Designing the BIS to create a uniform field around the target area ensures predictable deflection of bullets.

Analyzing the Force Exerted on Moving Electric Charges

The force exerted on moving electric charges in a magnetic field is governed by the Lorentz Force Law. This force plays a crucial role in the BIS's ability to deflect bullets.

Lorentz Force Law

The Lorentz Force $\mathbf{F}$ on a charged particle with charge q moving with velocity $\mathbf{v}$ in a magnetic field $\mathbf{B}$ is given by:

$$\mathbf{F} = q(\mathbf{v} \times \mathbf{B})$$

This equation highlights several important aspects:

1. **Direction of Force**: The force is perpendicular to both the velocity of the particle and the magnetic field, following the right-hand rule. This perpendicular force causes the particle to move in a circular or spiral path.
2. **Magnitude of Force**: The force's magnitude depends on the charge q, the velocity v, and the magnetic field strength B. Faster-moving particles or stronger magnetic fields result in greater forces.

Application to Bullet Deflection

To deflect a bullet, the BIS must generate a magnetic field that interacts with the bullet as it approaches the protected individual. Since most bullets are non-magnetic and electrically neutral, the system must induce a current in the bullet or use other methods to make the bullet susceptible to the magnetic field.

Induced Currents

Using Faraday's Law of Induction, a changing magnetic field can induce an electromotive force (EMF) in a conductive material. As a bullet travels through a magnetic field, the field changes relative to the bullet's motion, inducing currents within the bullet. These induced currents create their own magnetic fields, which then interact with the original magnetic field, producing a Lorentz force that alters the bullet's trajectory.

The induced EMF $\mathcal{E}$ in a bullet moving through a magnetic field is given by:

$$dt\mathcal{E} = -\frac{d\Phi_B}{dt}$$

where Φ_B is the magnetic flux through the bullet. The induced current creates a magnetic moment that interacts with the external magnetic field, resulting in a force that can deflect the bullet.

Practical Considerations

Designing a BIS to effectively use magnetic fields for bullet deflection involves several practical considerations:

1. **Magnetic Field Strength**: The magnetic field must be strong enough to induce significant currents in the bullet and produce a deflective force. This requires high-power electromagnets or superconducting materials capable of generating fields in the range of several teslas.
2. **Field Configuration**: The magnetic field must be configured to maximize interaction with the bullet. This involves designing the field's geometry to ensure the bullet passes through regions of high field strength and experiences a force sufficient to alter its trajectory.
3. **Bullet Composition**: Since most bullets are non-magnetic, materials or coatings that enhance their interaction with magnetic fields may be needed. Alternatively, the BIS could use a combination of electric and magnetic fields to induce currents in non-magnetic bullets.
4. **Energy Efficiency**: Generating and maintaining strong magnetic fields requires substantial energy. Efficient power sources, such as advanced batteries or supercapacitors, and energy-saving techniques, such as pulsed field generation, are essential for the BIS to be practical and portable.

Testing and Validation

Extensive testing and validation are critical to ensuring the BIS operates effectively and safely. This includes:

1. **Simulation**: Computer simulations can model the interaction between bullets and magnetic fields, allowing for the optimization of field strength and configuration without the need for physical prototypes.
2. **Prototype Development**: Building and testing physical prototypes to validate theoretical models and refine the system's design. This involves measuring the deflection of bullets under controlled conditions and assessing the system's reliability and robustness.
3. **Safety Considerations**: Ensuring that the BIS does not pose a risk to the user or bystanders. This includes addressing potential hazards associated with strong magnetic fields and ensuring the system operates predictably and fails safely.

Conclusion

The use of magnetic fields to deflect bullets is a promising and innovative approach to personal protection, leveraging fundamental principles of electromagnetism. Understanding the behavior of magnetic fields and the forces they exert on moving electric charges is crucial for developing effective bullet deflection systems like the BIS. By inducing currents in bullets and creating strong, well-configured magnetic fields, the BIS can alter bullet trajectories, offering a proactive and dynamic defense mechanism. However, significant practical challenges, including magnetic field strength, energy efficiency, and safety, must be addressed through rigorous research, development, and testing. With continued advances in technology and material science, the BIS has the potential to revolutionize personal protection, providing a new level of safety against ballistic threats.

Material Science for BIS

The development of a Bullet Inversion System (BIS) that can deflect or neutralize bullets using magnetic fields relies heavily on the principles of material science. Understanding the properties of ferromagnetic materials like iron, cobalt, and nickel, as well as addressing the challenges posed by lead and other non-magnetic materials commonly used in bullets, is essential for creating an effective BIS. This discussion will delve into the relevant properties of these materials and explore potential solutions to the challenges they present.

Properties of Ferromagnetic Materials: Iron, Cobalt, Nickel

Ferromagnetic materials are those that can be magnetized and exhibit strong interactions with magnetic fields. Among the most well-known ferromagnetic materials are iron, cobalt, and nickel. These materials possess unique properties that make them suitable for various applications in electromagnetism, including the potential use in a BIS.

Iron

Iron is one of the most abundant and widely used ferromagnetic materials. It has a high magnetic permeability, which means it can easily become magnetized and retain its magnetic properties. The key properties of iron include:

1. **High Magnetic Permeability**: Iron can be easily magnetized, and its high permeability allows it to support strong magnetic fields.
2. **Curie Temperature**: The Curie temperature of iron is about 770°C (1420°F). Above this temperature, iron loses its ferromagnetic properties and becomes paramagnetic.
3. **Abundance and Cost**: Iron is abundant and relatively inexpensive, making it an attractive material for large-scale applications.

Cobalt

Cobalt is another ferromagnetic material with properties that make it useful in high-temperature applications and where corrosion resistance is required. Key properties of cobalt include:

1. **High Magnetic Strength**: Cobalt exhibits strong magnetic properties, with a higher coercivity than iron, meaning it retains its magnetization better under external magnetic influences.
2. **Curie Temperature**: Cobalt has a high Curie temperature of around 1,115°C (2,039°F), which makes it suitable for applications requiring stability at high temperatures.
3. **Corrosion Resistance**: Cobalt is more resistant to oxidation and corrosion compared to iron, enhancing its durability in various environments.

Nickel

Nickel is known for its ferromagnetic properties and its ability to alloy with other metals to enhance their magnetic characteristics. Key properties of nickel include:

1. **Moderate Magnetic Strength**: Nickel has moderate magnetic strength, with a lower coercivity compared to cobalt but higher than iron.
2. **Curie Temperature**: Nickel's Curie temperature is about 358°C (676°F), making it suitable for applications within this temperature range.
3. **Alloying Capabilities**: Nickel is often alloyed with other metals to improve their magnetic properties and corrosion resistance.

Challenges Posed by Lead and Non-Magnetic Materials in Bullets

Most bullets are made from lead or lead-based alloys, which are non-magnetic. This poses a significant challenge for the BIS, as these materials do not interact directly with magnetic fields in the same way ferromagnetic materials do. Addressing this challenge requires innovative approaches and alternative solutions.

Lead Properties

Lead is chosen for bullets primarily due to its density and malleability, which contribute to its effectiveness as a projectile. However, lead's non-magnetic nature complicates the use of magnetic fields for deflection. Key properties of lead include:

1. **High Density**: Lead's high density gives bullets their penetrating power and stability in flight.
2. **Softness and Malleability**: Lead is easily shaped and deformed, which is useful in manufacturing and upon impact.
3. **Non-Magnetic Nature**: Lead does not respond to magnetic fields, making it difficult to manipulate using traditional magnetic methods.

Potential Solutions

To address the challenge posed by lead and non-magnetic materials, several approaches can be considered:

1. **Electromagnetic Induction**: One potential solution is to use electromagnetic induction to induce currents in the bullet. As the bullet passes through a magnetic field, a changing magnetic flux can induce an electromotive force (EMF) and generate currents within the bullet. These induced currents can create their own magnetic fields, which then interact with the external magnetic field to produce a deflective force.
2. **Material Coatings**: Coating bullets with ferromagnetic materials such as iron, cobalt, or nickel can make them more responsive to magnetic fields. This approach involves

adding a thin layer of a ferromagnetic material to the bullet's surface, allowing the BIS to exert a force on the bullet more effectively.

3. **Composite Materials**: Developing composite materials that combine the desirable properties of lead (such as density and malleability) with magnetic responsiveness can also be a viable solution. These composites could be engineered to maintain the ballistic performance of lead while being influenced by magnetic fields.

4. **High-Energy Plasma Shield**: Another advanced approach involves using a high-energy plasma shield to deflect bullets. Plasma, an ionized gas, can be manipulated using electromagnetic fields to create a barrier that can deflect or disrupt the trajectory of bullets. This method requires significant energy and sophisticated control systems but offers a potential solution for dealing with non-magnetic projectiles.

Practical Implementation

Implementing these solutions in a practical BIS involves several considerations:

1. **Power Supply**: Generating and maintaining the necessary magnetic fields or plasma shields requires a substantial and reliable power supply. Advances in battery technology, supercapacitors, and energy harvesting methods are critical to making the BIS portable and effective.

2. **Field Configuration**: The magnetic or electromagnetic fields must be carefully configured to maximize their interaction with the bullet. This involves designing the BIS to create a focused and intense field in the path of the bullet while minimizing energy consumption.

3. **Material Durability**: The materials used in the BIS, as well as any coatings or composites applied to bullets, must withstand the stresses of high-speed impacts and environmental factors. Ensuring durability and reliability is essential for the BIS to function effectively in real-world conditions.

4. **Safety and Regulation**: Any system designed to deflect bullets must adhere to strict safety standards and regulations. This includes ensuring that the BIS does not pose risks to the user or bystanders and that it operates within legal frameworks governing the use of such technologies.

Conclusion

Material science plays a pivotal role in the development of the Bullet Inversion System, with a focus on understanding the properties of ferromagnetic materials like iron, cobalt, and nickel, as well as addressing the challenges posed by lead and other non-magnetic materials. By leveraging electromagnetic induction, material coatings, composite materials, and advanced plasma technologies, it is possible to create a BIS capable of deflecting bullets. Practical implementation requires careful consideration of power supply, field configuration, material durability, and safety regulations. With continued research and innovation, the BIS has the potential to revolutionize personal protection by providing a dynamic and proactive defense against ballistic threats.

Electromagnetic Induction

Electromagnetic induction is a fundamental principle of electromagnetism that has wide-ranging applications in modern technology. It involves the generation of an electromotive force (EMF) across a conductor when it experiences a changing magnetic field. This concept, discovered by Michael Faraday in the 19th century, is encapsulated in Faraday's Law of Induction and has profound implications for the development of technologies like the Bullet Inversion System (BIS).

Faraday's Law of Induction and its Application

Faraday's Law of Induction states that the EMF induced in a circuit is directly proportional to the rate of change of the magnetic flux through the circuit. Mathematically, it is expressed as:

$$dt\mathcal{E} = -\frac{d\Phi_B}{dt}$$

where $\mathcal{E}$ is the induced EMF and Φ_B is the magnetic flux. The negative sign in the equation represents Lenz's Law, which indicates that the direction of the induced EMF opposes the change in magnetic flux that caused it.

Magnetic Flux

Magnetic flux Φ_B is defined as the product of the magnetic field B and the area A through which the field lines pass, taking into account the angle θ between the magnetic field and the normal to the surface:

$$\Phi_B = B \cdot A \cdot \cos(\theta)$$

In practical applications, changing the magnetic flux through a conductor can be achieved by varying the strength of the magnetic field, the area of the loop, or the orientation of the field with respect to the loop.

Inducing Currents in Bullets to Generate Magnetic Fields

One of the innovative applications of Faraday's Law of Induction is in the concept of a Bullet Inversion System (BIS), which aims to deflect or neutralize bullets using electromagnetic fields. Given that most bullets are made of non-magnetic materials like lead, directly manipulating them with magnetic fields is challenging. However, by inducing currents within the bullets as they travel through a magnetic field, it is possible to generate secondary magnetic fields that interact with the primary field, exerting forces that can alter the bullet's trajectory.

Induced Currents in Conductive Bullets

When a conductive bullet passes through a magnetic field, the changing magnetic flux induces an EMF within the bullet. This induced EMF drives currents through the bullet, creating what is known as eddy currents. These currents generate their own magnetic fields, which interact with the external magnetic field, leading to the Lorentz force that can deflect the bullet.

Mathematical Model

To understand this process quantitatively, consider a bullet of length LLL moving with velocity vvv through a magnetic field BBB. The rate of change of magnetic flux through the bullet is given by:

$$\frac{d\Phi_B}{dt} = B \cdot v \cdot L dt$$

According to Faraday's Law, the induced EMF E\mathcal{E}E is:

$$\mathcal{E} = -B \cdot v \cdot L$$

This EMF drives a current III through the bullet, which in turn generates a magnetic field that opposes the change in flux, as described by Lenz's Law. The interaction between this induced magnetic field and the external field creates a force on the bullet.

Lorentz Force

The Lorentz force $\mathbf{F}$ acting on a segment of the bullet with induced current III in the presence of a magnetic field B is given by:

$$\mathbf{F} = I \cdot L \times B$$

The direction of this force is perpendicular to both the direction of the current and the magnetic field, according to the right-hand rule. This force can be harnessed to deflect the bullet away from its original path.

Practical Considerations for BIS Implementation

Implementing electromagnetic induction to deflect bullets involves several practical considerations and technological challenges.

Magnetic Field Generation

The BIS must generate a strong and well-controlled magnetic field to induce sufficient currents in the bullet. This requires powerful electromagnets or superconducting magnets capable of producing fields in the range of several teslas. The field must also be configured to maximize its interaction with the bullet, ensuring that the bullet experiences a significant change in magnetic flux as it passes through the field.

Energy Requirements

Generating and maintaining strong magnetic fields requires substantial energy. Advances in battery technology and supercapacitors are critical to providing the necessary power in a portable format. Efficient energy management systems are also needed to optimize the power usage and ensure the BIS can operate effectively over extended periods.

Material Considerations

Since most bullets are made of lead or other non-magnetic materials, inducing currents within them can be challenging. Coating bullets with a thin layer of conductive material or developing composite bullets that combine lead with conductive components can enhance their responsiveness to magnetic fields. Alternatively, high-energy plasma or other advanced materials could be explored to create barriers that interact with non-magnetic bullets.

Field Configuration and Control

The magnetic field generated by the BIS must be precisely controlled to ensure it interacts effectively with the bullet. This involves designing the field's geometry and strength to create a focused area where the bullet experiences maximum deflection. Advanced control systems are needed to dynamically adjust the field in response to the bullet's trajectory and speed.

Safety and Testing

Safety is paramount in developing the BIS. The system must be rigorously tested to ensure it does not pose risks to the user or bystanders. This includes assessing the effects of strong magnetic fields on electronic devices and ensuring the system operates predictably and fails safely. Extensive testing, both in simulation and with physical prototypes, is essential to validate the BIS's effectiveness and reliability.

Future Research and Development

Continued research and development are crucial to advancing the BIS concept. This includes exploring new materials and coatings that enhance the magnetic responsiveness of bullets, developing more efficient energy storage and management systems, and refining the control mechanisms for generating and directing magnetic fields. Collaboration between material scientists, engineers, and physicists will be essential to overcoming the technical challenges and realizing the potential of the BIS.

Advanced Materials

Research into advanced materials that exhibit superior magnetic and conductive properties could lead to significant improvements in the BIS. This includes exploring novel alloys, composites, and nanomaterials that can be used in bullet coatings or as part of the BIS's magnetic field generation system.

Energy Efficiency

Improving the energy efficiency of the BIS is critical for its practicality. This involves not only advances in battery technology and supercapacitors but also the development of energy-harvesting techniques and power management algorithms that optimize the system's performance.

Field Dynamics

Understanding and controlling the dynamics of the magnetic fields generated by the BIS is essential for effective bullet deflection. This includes developing algorithms and control systems that can adapt the field configuration in real-time, based on the bullet's speed, direction, and material properties.

Conclusion

Electromagnetic induction, as described by Faraday's Law, offers a promising avenue for the development of the Bullet Inversion System. By inducing currents in bullets and generating magnetic fields, the BIS can exert forces that alter the trajectory of incoming projectiles. Practical implementation requires careful consideration of magnetic field generation, energy requirements, material properties, and field control. Continued research and innovation are essential to overcome these challenges and realize the potential of the BIS, offering a new level of personal protection against ballistic threats.

Plasma Physics for Bullet Deflection

Plasma physics represents a frontier in advanced defense technologies, particularly in the context of deflecting bullets and other projectiles. Plasma, often referred to as the fourth state of matter, consists of an ionized gas with unique properties that can be harnessed for creating protective barriers. This exploration delves into the properties of plasma, the challenges associated with creating and maintaining a high-energy plasma shield, and the potential applications of such technology in bullet deflection.

Plasma as an Ionized Gas and Its Properties

Plasma is a state of matter in which the gas phase is energized until atomic electrons are no longer associated with any particular atomic nucleus. This results in a mixture of free electrons and ions, giving plasma its distinctive properties. Understanding these properties is crucial for leveraging plasma in bullet deflection systems.

Ionization

Ionization is the process of adding or removing electrons from an atom or molecule, thereby creating ions. Plasma can be generated by heating a gas to high temperatures or by applying a strong electromagnetic field. The degree of ionization depends on the energy input, with higher energy levels producing a greater proportion of ions and free electrons.

Conductivity

Plasma is an excellent conductor of electricity due to the presence of free electrons. This high conductivity allows plasma to interact strongly with electromagnetic fields, which can be used to control and stabilize the plasma.

Magnetic Fields

Plasma can be influenced and confined by magnetic fields. This property is pivotal in creating a stable plasma shield, as magnetic fields can be used to contain the ionized gas and shape it into a protective barrier.

Temperature and Pressure

Plasmas typically exist at high temperatures and can generate significant pressure. These conditions pose challenges for containment and stability but also provide the potential for creating a powerful deflective barrier.

Creating and Maintaining a High-Energy Plasma Shield

The concept of a high-energy plasma shield for bullet deflection involves generating and sustaining a plasma field that can absorb, deflect, or disrupt incoming projectiles. This requires sophisticated technology and careful management of the plasma's properties.

Generating Plasma

Generating plasma for a shield involves ionizing a gas such as argon, neon, or even air. There are several methods to achieve this:

1. **Thermal Ionization**: Heating a gas to extremely high temperatures can ionize it, creating plasma. This method requires significant energy input but is straightforward in principle.
2. **Electromagnetic Ionization**: Using strong electric or magnetic fields to ionize the gas. Techniques such as inductively coupled plasma (ICP) or microwave ionization can be employed to generate plasma efficiently.
3. **Laser Ionization**: High-powered lasers can ionize gas particles along their path, creating a plasma channel. This method offers precision and control but also requires advanced laser technology.

Maintaining Plasma

Once generated, the plasma must be maintained and stabilized to form an effective shield. This involves several key components:

1. **Magnetic Confinement**: Using magnetic fields to contain and shape the plasma. This can be achieved with devices such as magnetic mirrors, tokamaks, or stellarators, which use complex magnetic field configurations to confine plasma.
2. **Energy Input**: Continuously supplying energy to the plasma to maintain its ionization and temperature. This can be done through electrical currents, electromagnetic waves, or other energy sources.
3. **Cooling and Stability**: Managing the heat generated by the plasma to prevent destabilization or damage to surrounding structures. Advanced cooling systems and materials capable of withstanding high temperatures are essential.

Field Configuration

The configuration of the magnetic and electric fields used to contain the plasma is critical for the effectiveness of the shield. The fields must be designed to create a stable and uniform plasma barrier, capable of interacting with and deflecting incoming bullets.

Practical Applications and Challenges

Implementing a plasma shield for bullet deflection presents several practical challenges and potential solutions.

Energy Requirements

The energy required to generate and maintain a high-energy plasma shield is substantial. Developing efficient energy sources and storage systems, such as advanced batteries, supercapacitors, or compact fusion reactors, is essential for making the technology viable.

Material Constraints

Materials used in the construction of the plasma generation and confinement system must withstand extreme temperatures and electromagnetic fields. Research into high-temperature superconductors, advanced ceramics, and other resilient materials is crucial.

Control Systems

Advanced control systems are needed to manage the plasma's behavior, ensuring it remains stable and effective as a deflective barrier. This involves real-time monitoring and adjustments of the magnetic and electric fields, as well as the energy input.

Safety Considerations

Safety is paramount when dealing with high-energy plasma. The system must be designed to minimize risks to users and bystanders, with fail-safes and containment measures in place to prevent accidental exposure or malfunction.

Testing and Validation

Extensive testing and validation are required to ensure the plasma shield operates effectively and reliably under various conditions. This includes laboratory experiments, computer simulations, and field tests to evaluate performance and identify potential issues.

Future Research and Development

Continued research and development are essential to advance plasma shield technology. Key areas of focus include:

Advanced Plasma Generation

Developing new methods for efficient and controlled plasma generation. This includes exploring alternative ionization techniques and optimizing existing methods for higher efficiency and scalability.

Energy Efficiency

Improving the energy efficiency of plasma generation and maintenance. This involves not only advances in energy sources and storage but also optimizing the energy use within the plasma system itself.

Material Innovations

Researching and developing new materials capable of withstanding the extreme conditions associated with plasma confinement and stability. This includes high-temperature superconductors, advanced composites, and other resilient materials.

Field Dynamics

Enhancing the understanding and control of magnetic and electric fields used to contain and shape plasma. This includes developing more sophisticated algorithms and control systems to dynamically adjust field configurations for optimal performance.

Integration with Other Technologies

Exploring the integration of plasma shields with other defense technologies, such as advanced body armor, automated threat detection systems, and energy weapons. This holistic approach can create more comprehensive and effective personal protection systems.

Conclusion

Plasma physics offers a promising avenue for developing advanced bullet deflection systems through the use of high-energy plasma shields. Understanding the properties of plasma as an ionized gas and addressing the challenges of generating and maintaining a stable plasma barrier are critical to realizing this technology. Practical implementation requires significant advancements in energy efficiency, material science, and control systems. With continued research and innovation, plasma shields have the potential to revolutionize personal protection by providing a dynamic and powerful defense against ballistic threats.

Power Requirements and Solutions

In developing advanced protective technologies such as the Bullet Inversion System (BIS) or a high-energy plasma shield, one of the most critical aspects is addressing the power requirements. Generating and maintaining strong magnetic fields or high-energy plasma demands significant power, necessitating innovative solutions in energy storage and management. This discussion explores the power needs for such systems, focusing on estimating the required power and examining high-energy batteries and supercapacitors as potential solutions.

Estimating Power Needs for Generating Strong Magnetic Fields

Creating strong magnetic fields, necessary for technologies like the BIS, involves substantial power consumption. To estimate the power requirements, we must consider the fundamental principles of electromagnetism and the specific characteristics of the intended application.

Magnetic Field Strength

The strength of the magnetic field, denoted as B, is a key factor. For effective bullet deflection, the magnetic field must be strong enough to exert significant forces on the projectiles. The required field strength depends on the bullet's speed, mass, and material properties. Fields in the range of several teslas (T) are typically needed for practical deflection.

Electromagnet Design

The design of the electromagnet plays a crucial role in determining power needs. The magnetic field B produced by an electromagnet is given by:

$$LB = \mu_0 \cdot \frac{N \cdot I}{L}$$

where μ_0 is the permeability of free space, N is the number of turns in the coil, I is the current, and L is the length of the coil. Achieving high field strengths requires high current and/or a large number of turns, both of which increase power consumption.

Power Consumption

The power P required to generate the magnetic field can be estimated using the formula:

$$P = I^2 \cdot R$$

where R is the resistance of the coil. High currents lead to significant power dissipation as heat, necessitating efficient cooling systems to manage thermal loads.

Energy Storage Needs

To sustain the magnetic field, the system requires a continuous power supply. The total energy E needed over a period ttt can be calculated as:

$$E = P \cdot t$$

For portable applications like personal protection, this energy must be stored in compact and efficient batteries or capacitors.

Exploring High-Energy Batteries and Supercapacitors

Meeting the power requirements of advanced magnetic field generation and plasma maintenance involves leveraging state-of-the-art energy storage technologies. High-energy batteries and supercapacitors offer promising solutions, each with distinct advantages and challenges.

High-Energy Batteries

High-energy batteries are designed to store large amounts of energy in a compact form. They are essential for providing the sustained power needed for systems like the BIS or plasma shields.

1. **Lithium-Ion Batteries**

Lithium-ion (Li-ion) batteries are the most widely used high-energy storage devices. They offer high energy density, long cycle life, and relatively low self-discharge rates. Advances in Li-ion technology continue to improve their performance, making them suitable for high-power applications.

2. **Solid-State Batteries**

Solid-state batteries represent the next generation of energy storage, replacing the liquid electrolyte in traditional Li-ion batteries with a solid electrolyte. This change enhances safety, increases energy density, and allows for faster charging and discharging rates. Solid-state batteries are particularly promising for applications requiring high power and energy efficiency.

3. **Metal-Air Batteries**

Metal-air batteries, such as lithium-air or zinc-air batteries, offer extremely high energy densities by utilizing oxygen from the air as a reactant. These batteries are still in the experimental stages but have the potential to revolutionize energy storage with their lightweight and high-capacity characteristics.

Supercapacitors

Supercapacitors, also known as ultracapacitors, store energy through electrostatic separation of charges rather than chemical reactions. This allows for rapid charging and discharging, making them ideal for applications requiring quick bursts of power.

1. **High Power Density**

Supercapacitors have much higher power densities compared to batteries, meaning they can deliver large amounts of power in short time periods. This makes them suitable for applications like magnetic field pulsing in BIS or initiating plasma generation.

2. **Longevity and Efficiency**

Supercapacitors can withstand millions of charge-discharge cycles without significant degradation, offering superior longevity and efficiency. This reduces maintenance and replacement costs over the system's lifespan.

3. **Hybrid Energy Systems**

Combining supercapacitors with high-energy batteries creates a hybrid energy storage system that leverages the strengths of both technologies. Batteries provide sustained energy, while supercapacitors handle peak power demands. Such hybrid systems optimize performance, efficiency, and longevity.

Practical Implementation and Future Directions

Implementing high-energy batteries and supercapacitors in advanced protection systems involves addressing several practical challenges, including energy management, thermal regulation, and integration with other system components.

Energy Management Systems

Effective energy management systems (EMS) are crucial for optimizing the performance of high-energy batteries and supercapacitors. An EMS monitors and controls the energy flow, ensuring efficient use of stored power and preventing overcharging or deep discharging, which can degrade the storage devices.

Thermal Regulation

Both high-energy batteries and supercapacitors generate heat during operation, especially under high power loads. Efficient thermal regulation systems are needed to dissipate this heat, maintaining optimal operating temperatures and preventing thermal runaway. Advanced cooling techniques, such as liquid cooling or phase change materials, can enhance thermal management.

Integration and Miniaturization

For personal protection applications, the energy storage system must be compact and lightweight. Integrating high-energy batteries and supercapacitors into a portable device requires careful design to minimize size and weight while maximizing energy capacity and power output. Advances in materials science and nanotechnology can contribute to the miniaturization and integration of these components.

Future Research and Development

Ongoing research and development are essential to push the boundaries of energy storage technology. Key areas of focus include:

1. **Advanced Materials**

Exploring new materials with higher energy densities, better thermal properties, and longer lifespans can significantly enhance the performance of batteries and supercapacitors. This includes research into novel electrode materials, solid electrolytes, and nanostructured components.

2. **Improved Manufacturing Techniques**

Developing more efficient and cost-effective manufacturing techniques can make advanced energy storage technologies more accessible and scalable. Innovations in additive manufacturing, such as 3D printing, can enable the precise fabrication of complex battery and supercapacitor designs.

3. **Energy Harvesting**

Integrating energy harvesting technologies with energy storage systems can provide additional power sources, extending the operational time of protection devices. This includes harvesting ambient energy from solar, thermal, or kinetic sources.

4. **System Optimization**

Optimizing the overall system design, including power electronics, control algorithms, and user interfaces, can enhance the efficiency and usability of protection technologies. Advanced simulation and modeling tools can aid in the design and testing of these integrated systems.

Conclusion

Estimating and meeting the power requirements for advanced protective technologies like the Bullet Inversion System and high-energy plasma shields is a complex challenge. High-energy batteries and supercapacitors offer promising solutions, each with unique advantages and technical hurdles. By leveraging advances in materials science, energy management, and system integration, it is possible to develop efficient, compact, and reliable energy storage systems that enable the practical implementation of these cutting-edge technologies. Continued research and innovation are essential to overcome the remaining challenges and realize the full potential of these advanced protective systems.

Miniaturization and Portability

Miniaturization and portability are pivotal in the advancement of modern technology, particularly in fields such as personal protection systems. For applications like the Bullet Inversion System (BIS) and high-energy plasma shields, making these technologies compact and portable involves overcoming significant technical challenges. This discussion delves into the complexities of miniaturizing magnetic field generators and plasma systems, and outlines the principles and strategies for designing a portable BIS device.

Challenges in Miniaturizing Magnetic Field Generators and Plasma Systems

1. Magnetic Field Generators

Magnetic field generators, essential for technologies like the BIS, require precise engineering to produce strong magnetic fields in a compact form. The primary challenges include:

- **Field Strength and Size**: The strength of a magnetic field is directly proportional to the current flowing through the coils and inversely proportional to the coil's length. To generate strong fields, high currents are necessary, which traditionally implies larger coil dimensions. Miniaturizing the coils while maintaining or enhancing field strength demands innovative solutions in coil design and material science.

- **Heat Dissipation**: High currents generate significant heat, which must be managed to prevent overheating and maintain performance. Miniaturization constrains the available space for cooling systems. Advanced cooling techniques, such as micro-channel heat exchangers or phase-change materials, are required to efficiently dissipate heat in compact designs.

- **Material Constraints**: Superconducting materials can carry high currents with minimal resistive losses, but they are expensive and complex to handle. Research into high-temperature superconductors and their integration into miniaturized systems is crucial for overcoming material constraints.

2. Plasma Systems

Creating and maintaining a high-energy plasma shield in a portable form presents distinct challenges:

- **Plasma Generation**: Plasma generation requires substantial energy to ionize a gas. Compact plasma generation devices must balance power density with space constraints. Techniques like microwave or laser ionization offer potential solutions, but integrating these technologies into a portable format involves advanced engineering.

- **Magnetic Confinement**: Plasma confinement relies on magnetic fields to stabilize and shape the plasma. Designing miniaturized magnetic confinement systems requires precision in field generation and control. Miniaturized magnetic confinement systems must integrate high-field-strength magnets or advanced magnetic materials without sacrificing stability.

- **Energy Management**: Plasma systems consume significant amounts of power. Efficient energy storage and management are essential for portable systems. Innovations in energy storage, such as high-energy batteries and supercapacitors, need to be compact and capable of delivering rapid bursts of power.

Designing a Portable BIS Device

Designing a portable Bullet Inversion System (BIS) involves several key considerations, including size, weight, power management, and user interface.

1. Size and Weight

The physical dimensions and weight of the BIS device must be minimized without compromising functionality. This involves:

- **Compact Design**: Employing miniaturization techniques to reduce the size of magnetic field generators and energy storage components. Advanced materials and manufacturing techniques, such as additive manufacturing or micro-fabrication, can enable the creation of compact and efficient components.
- **Lightweight Materials**: Using lightweight, high-strength materials, such as carbon fiber composites or advanced polymers, to reduce the overall weight of the device. These materials must balance strength, weight, and cost while ensuring durability.

2. Power Management

Efficient power management is critical for portable BIS devices. This involves:

- **Energy Storage**: Incorporating high-energy batteries or supercapacitors that are compact and capable of providing sufficient power. Battery technologies such as solid-state batteries or advanced lithium-ion cells offer high energy densities and compact form factors.

- **Power Electronics**: Designing efficient power electronics to manage the distribution and conversion of energy. This includes high-efficiency converters, regulators, and controllers that can operate within the compact constraints of the device.

- **Thermal Management**: Implementing effective thermal management solutions to handle heat generated by high-power components. Techniques such as heat sinks, thermal pads, or micro-channel cooling can help dissipate heat while maintaining a small footprint.

3. User Interface

A portable BIS device must be user-friendly and intuitive. This involves:

- **Control Systems**: Designing an interface for users to control and monitor the BIS device. This can include touchscreens, buttons, or voice commands. The interface must provide real-time feedback on system status, power levels, and operational modes.

- **Safety Features**: Incorporating safety mechanisms to protect users from potential hazards. This includes fail-safes, emergency shutdown systems, and user alerts. Ensuring that these features are integrated into the device without adding excessive bulk is crucial.

- **Ergonomics**: Designing the device to be comfortable and easy to handle. This includes considering the device's weight distribution, grip design, and overall user experience to ensure that it can be used effectively in various scenarios.

4. Integration and Modularity

A portable BIS device must be designed with integration and modularity in mind. This involves:

- **Modular Components**: Designing the device with modular components that can be easily replaced or upgraded. This allows for flexibility in improving or repairing the device without requiring complete redesigns.
- **Integration with Other Technologies**: Considering how the BIS device can integrate with other personal protection technologies. This may include compatibility with advanced body armor, threat detection systems, or communication devices.

5. Testing and Validation

Rigorous testing and validation are essential to ensure that the portable BIS device functions as intended. This includes:

- **Prototype Testing**: Developing and testing prototypes to evaluate performance, durability, and user experience. Field tests in various environments can help identify potential issues and refine the design.
- **Compliance and Certification**: Ensuring that the device meets relevant safety, performance, and regulatory standards. This may involve obtaining certifications from regulatory bodies or industry organizations.

Future Directions

Looking ahead, several areas offer potential for advancing the miniaturization and portability of BIS devices:

- **Nanotechnology**: Leveraging nanotechnology to develop smaller, more efficient components. Nanomaterials and nanostructures can enhance performance and reduce size, contributing to more compact and powerful devices.

- **Advanced Manufacturing**: Utilizing advanced manufacturing techniques, such as 3D printing and micro-fabrication, to create intricate and miniaturized components. These techniques can enable the precise fabrication of small-scale magnetic field generators and plasma systems.

- **Integration of AI**: Incorporating artificial intelligence for enhanced control and optimization of the BIS device. AI can assist in real-time adjustments, predictive maintenance, and adaptive responses to threats.

Conclusion

Miniaturization and portability are critical challenges in developing advanced protection technologies like the Bullet Inversion System and plasma shields. Overcoming these challenges involves innovative approaches to component design, power management, and user interface development. By leveraging advanced materials, energy storage technologies, and manufacturing techniques, it is possible to create compact and effective personal protection devices. Continued research and development in these areas will be essential for realizing the full potential of portable BIS devices and other advanced defense technologies.

Magnetic Field Generators

Magnetic field generators are pivotal components in various technologies, ranging from industrial applications to advanced personal protection systems. Their design and functionality depend heavily on the type of magnet used and the materials employed. This discussion explores the types and designs of electromagnets and permanent magnets, as well as the utilization of superconducting materials for compact, high-performance magnetic field generators.

Types and Designs of Electromagnets and Permanent Magnets

1. Electromagnets

Electromagnets are devices that generate a magnetic field through the flow of electric current. They are widely used due to their ability to produce adjustable magnetic fields. The fundamental design aspects of electromagnets include:

- **Coil Design**: The core component of an electromagnet is the coil, typically wound around a core material. The magnetic field strength B is determined by the current III flowing through the coil, the number of turns N, and the core material. The relationship is given by:

$$LB = \mu \cdot \frac{N \cdot I}{L}$$

where μ\muμ is the magnetic permeability of the core material, and L is the length of the coil. Increasing the number of turns or the current enhances the field strength but requires careful design to manage power and heat dissipation.

- **Core Materials**: The choice of core material affects the efficiency of the electromagnet. Soft ferromagnetic materials, such as iron, are commonly used due to their high magnetic permeability, which concentrates the magnetic field lines and improves the field strength. Advanced materials, such as silicon steel or laminated cores, are used in applications requiring high performance and reduced energy losses.
- **Cooling Systems**: High-current electromagnets generate substantial heat. Effective cooling systems, such as air cooling or liquid cooling, are essential to maintain operational stability and prevent overheating. Advanced designs incorporate heat sinks and cooling channels to manage thermal loads efficiently.

2. Permanent Magnets

Permanent magnets generate a magnetic field without the need for an external power source. Their magnetic field is due to the intrinsic properties of the material, which retains its magnetization over time. Key types and designs of permanent magnets include:

- **Neodymium Magnets**: Neodymium-iron-boron (NdFeB) magnets are among the strongest permanent magnets available. They are composed of an alloy of neodymium, iron, and boron. These magnets are used in various applications due to their high magnetic strength and compact size. They are, however, sensitive to high temperatures and require protective coatings to prevent corrosion.

- **Samarium-Cobalt Magnets**: Samarium-cobalt (SmCo) magnets are another type of high-performance permanent magnet. They offer high magnetic strength and excellent thermal stability, making them suitable for high-temperature environments. Their inherent resistance to oxidation provides durability in harsh conditions.

- **Ceramic Magnets**: Ceramic or ferrite magnets are composed of iron oxide and barium or strontium carbonate. They are less expensive than rare-earth magnets and offer good magnetic performance for various applications. Their lower magnetic strength compared to neodymium or samarium-cobalt magnets limits their use in high-performance applications.

Utilizing Superconducting Materials for Compact Design

1. Superconductivity and Magnetic Fields

Superconductors are materials that, below a certain critical temperature, exhibit zero electrical resistance and perfect diamagnetism, known as the Meissner effect. This property allows superconductors to generate very strong magnetic fields without energy losses. Key points regarding superconducting materials include:

- **Critical Temperature**: Superconductors have a critical temperature (Tc) below which they transition to the superconducting state. High-temperature superconductors (HTS), such as yttrium barium copper oxide (YBCO), operate at relatively higher temperatures (around 90 K) compared to low-temperature superconductors (LTS) like niobium-titanium (NbTi), which require temperatures close to absolute zero.
- **Magnetic Field Strength**: Superconductors can produce magnetic fields much stronger than conventional electromagnets. The maximum field strength is limited by the material's critical magnetic field, which varies with temperature. For HTS materials, fields exceeding 30 teslas are achievable, making them ideal for high-performance applications.

2. Compact Design Considerations

Using superconducting materials allows for the creation of compact and high-performance magnetic field generators. Key considerations include:

- **Cooling Requirements**: Superconducting materials require cooling to maintain their superconducting state. HTS materials typically use liquid nitrogen as a coolant, which is more manageable and cost-effective compared to the liquid helium required for LTS materials. Compact cooling systems are essential to fit within the constrained dimensions of portable devices.

- **Cryostats**: A cryostat is a device used to maintain the low temperatures required for superconductors. Modern cryostats are designed to be compact and efficient, using insulation and advanced cooling techniques to minimize heat ingress. For portable applications, cryostats must be optimized for size and weight while ensuring effective cooling.

- **Magnet Design**: Superconducting magnets can be designed as solenoids or toroids, depending on the application. Solenoids involve winding superconducting wire into coils to generate a magnetic field along the axis. Toroids use circular coils to create a magnetic field within a closed loop. Each design has specific advantages and applications, such as compact field generation or uniform field distribution.

3. Integration and Applications

Integrating superconducting magnets into compact systems involves several key aspects:

- **Power Supply**: Superconducting magnets require a stable and reliable power supply for quenching (transitioning from superconducting to normal state) and cooling systems. High-efficiency power electronics and energy management systems are essential for maintaining optimal performance and safety.
- **Applications**: Superconducting magnets are used in various advanced technologies, including magnetic resonance imaging (MRI), particle accelerators, and fusion reactors. Their use in portable personal protection systems could revolutionize the field by providing compact, high-performance magnetic field generators.

Future Directions

The future of magnetic field generators lies in continued advancements in materials science, manufacturing techniques, and system integration:

- **Material Innovations**: Research into new superconducting materials and improved core materials for electromagnets can enhance performance and reduce costs. Advances in high-temperature superconductors and novel magnetic materials hold promise for further miniaturization and increased efficiency.

- **Manufacturing Techniques**: Innovations in manufacturing, such as precision fabrication and additive manufacturing, can enable the creation of complex and compact magnetic field generators. These techniques allow for the production of intricate designs and the integration of advanced materials.

- **Integration with Other Technologies**: Combining magnetic field generators with other advanced technologies, such as AI-driven control systems and advanced cooling techniques, can enhance performance and functionality. Integrating these systems into portable applications will require interdisciplinary research and development.

Conclusion

Magnetic field generators, including electromagnets, permanent magnets, and superconducting materials, play a crucial role in advancing technology. Understanding the types and designs of these magnets, along with the challenges and opportunities associated with superconductors, is essential for developing compact, high-performance systems. Continued research and innovation in materials science and manufacturing will drive the future of magnetic field generation, enabling new applications and improving existing technologies.

Plasma Generation Techniques

Plasma, the fourth state of matter, consists of a collection of ions, electrons, and neutral particles. Its unique properties, such as electrical conductivity and responsiveness to magnetic fields, make it useful in various applications from fusion research to advanced manufacturing. This discussion explores two fundamental plasma generation techniques: ionizing gases using microwave generators and employing magnetic confinement for stable plasma fields.

Ionizing Gases Using Microwave Generators

1. Microwave Plasma Generation

Microwave generators are pivotal in creating plasma by ionizing gases. The process involves several key steps and principles:

- **Microwave Interaction with Gases**: Microwaves are a form of electromagnetic radiation with frequencies typically ranging from 1 GHz to 300 GHz. When microwaves interact with a gas, they transfer energy to the gas molecules, raising their thermal energy and causing ionization. The microwave energy ionizes the gas by exciting electrons to high energy levels, where they escape from their atoms, creating free electrons and ions.

- **Dielectric Heating**: Microwaves heat gases through dielectric heating. This occurs when the electric field component of microwaves causes polar gas molecules to rotate rapidly. This rotational movement generates collisions between molecules, leading to a rise in temperature and eventually ionization. The efficiency of dielectric heating depends on the gas type and its dielectric properties.

- **Microwave Plasma Devices**: Devices such as microwave discharge tubes or plasma reactors are used to generate plasma. These devices typically consist of a microwave source, such as a magnetron or a klystron, and a resonant cavity that ensures the microwave energy is effectively coupled into the gas. The design of the resonant cavity is crucial for achieving uniform microwave distribution and efficient plasma generation.

- **Control Parameters**: Several parameters influence microwave plasma generation, including microwave power, frequency, pressure, and the type of gas. High microwave power can produce higher plasma densities, while lower frequencies and pressures may require longer interaction times to achieve sufficient ionization. Additionally, specific gases, such as argon or helium, are often used due to their favorable ionization characteristics.

2. Applications of Microwave-Plasma Techniques

Microwave-generated plasma finds applications in various fields:

- **Material Processing**: Microwave plasma is used in material processing, including etching and deposition in semiconductor manufacturing. The high energy of the plasma assists in the precise removal or deposition of materials on substrates.

- **Environmental Control**: Plasma generated by microwaves can be employed in environmental control systems for waste treatment and pollution control. Plasma technologies can effectively decompose hazardous substances and neutralize pollutants.

- **Fusion Research**: In fusion research, microwave plasma generation is used to study and achieve the conditions necessary for nuclear fusion. The ability to ionize gases and control plasma behavior is crucial for sustaining high-temperature plasmas needed for fusion reactions.

Magnetic Confinement for Stable Plasma Fields

1. Principles of Magnetic Confinement

Magnetic confinement is a technique used to stabilize and control plasma by using magnetic fields. The primary principles include:

- **Lorentz Force**: Magnetic confinement relies on the Lorentz force, which acts on charged particles moving through a magnetic field. The force, given by) $\mathbf{F} = q(\mathbf{v} \times \mathbf{B})$, where q is the charge of the particle, $\mathbf{v}$ is its velocity, and $\mathbf{B}$ is the magnetic field, causes the particles to spiral along the field lines. This spiraling motion helps to confine the plasma within a specific region.

- **Magnetic Field Geometry**: Different geometries of magnetic fields are used to confine plasma effectively. Common configurations include:

  - **Toroidal Confinement**: In toroidal confinement, plasma is shaped into a doughnut-like structure. Devices such as tokamaks and stellarators utilize this geometry to create a magnetic field that circulates around the plasma, providing stability and containment. The toroidal magnetic field is combined with a poloidal field (a field looping around the torus) to achieve effective confinement.

  - **Spherical Confinement**: Spherical confinement devices, such as the Polywell and the Field-Reversed Configuration (FRC), use magnetic fields that encase the plasma in a spherical or near-spherical shape. This geometry can help achieve stable confinement with potentially simpler designs compared to toroidal configurations.

- **Magnetic Field Strength**: The strength of the magnetic field is critical for effective confinement. Stronger magnetic fields can better contain the plasma and reduce losses due to interactions with the vessel walls. The strength is determined by the design of the magnetic coils and the power supply systems used to generate the field.

2. Challenges and Innovations in Magnetic Confinement

Several challenges and innovations in magnetic confinement systems are pivotal for advancing plasma technologies:

- **Confinement Time**: Achieving a sufficiently long confinement time is essential for maintaining stable plasma. This involves minimizing losses due to magnetic field imperfections, plasma turbulence, and interactions with the vessel walls. Techniques such as advanced magnetic coil designs and active control systems are employed to enhance confinement times.

- **Energy Input and Efficiency**: Efficiently supplying energy to the plasma while minimizing losses is crucial. Innovations in power supplies, such as high-frequency heating systems (e.g., radiofrequency heating) and improved superconducting magnets, are employed to increase energy efficiency and maintain plasma stability.

- **Material Challenges**: The interaction of plasma with the confinement vessel can lead to material degradation due to high temperatures and radiation. Research into advanced materials and coatings that can withstand these harsh conditions is ongoing to improve the durability and performance of confinement devices.

3. Applications of Magnetic Confinement

Magnetic confinement is used in several advanced technologies and research areas:

- **Nuclear Fusion**: The primary application of magnetic confinement is in nuclear fusion research. Devices like tokamaks and stellarators aim to achieve the conditions necessary for sustained fusion reactions, which have the potential to provide a virtually limitless and clean energy source.

- **Plasma Processing**: Magnetic confinement is also used in plasma processing applications, including surface treatment, coating, and modification of materials. The precise control of plasma conditions allows for enhanced processing capabilities and improved material properties.

- **Space Propulsion**: In space propulsion research, magnetic confinement is explored for advanced propulsion systems, such as magnetic sails or plasma thrusters. These systems aim to harness the benefits of plasma for efficient and high-speed space travel.

Conclusion

Plasma generation techniques, including microwave ionization and magnetic confinement, play a crucial role in advancing technologies and scientific research. By understanding and refining these techniques, we can enhance plasma applications across various fields, from industrial processing to fusion energy. Continued innovation and research in these areas will drive the development of more efficient and effective plasma systems, contributing to significant technological and scientific advancements.

Simulation of Bullet Inversion System (BIS)

Simulating the Bullet Inversion System (BIS) involves using advanced computer models to test various design configurations and predict system performance. This process is crucial for understanding the feasibility of a BIS and optimizing its design before physical prototypes are built. The simulation process encompasses developing accurate models, analyzing simulation data, and refining the BIS design based on insights gained from these analyses.

Using Computer Models to Test Different Configurations

1. Developing Accurate Computational Models

Creating effective simulations for the BIS requires the development of detailed and accurate computational models. These models must accurately represent the physical principles and components of the system:

- **Electromagnetic Field Modeling**: The BIS relies on generating strong magnetic fields to deflect bullets. Simulating these fields involves solving Maxwell's equations, which describe how electric and magnetic fields interact. Advanced numerical methods, such as finite element analysis (FEA) or finite difference time domain (FDTD) methods, are used to model these fields. The accuracy of the simulation depends on the precision of the model parameters, such as material properties, field strengths, and coil configurations.

- **Bullet Dynamics Modeling**: The behavior of a bullet in response to magnetic fields must be modeled accurately. This includes simulating the bullet's trajectory, velocity, and interaction with the magnetic fields generated by the BIS. Computational fluid dynamics (CFD) and particle tracking simulations are used to predict how the bullet's motion is affected by the applied magnetic forces.

- **Plasma Modeling**: If the BIS design includes plasma shielding, the simulation must incorporate plasma physics. This involves modeling ionized gas behavior, including the interaction of plasma with magnetic fields and its effect on the bullet's trajectory. Plasma models often use magnetohydrodynamics (MHD) to simulate the complex interactions between the plasma and electromagnetic fields.

2. Testing Design Variations

Once the computational models are established, various design configurations can be tested virtually:

- **Magnetic Field Configurations**: Different configurations of magnetic field generators, such as varying the number of coil turns, coil placement, and core materials, can be simulated to determine their impact on field strength and distribution. The goal is to identify the optimal configuration that provides effective bullet deflection while minimizing energy consumption and heat generation.

- **Power Supply Systems**: The efficiency and stability of power supply systems are critical for maintaining magnetic field strength. Simulations can test various power supply designs, including different types of batteries or supercapacitors, to assess their performance under dynamic conditions.

- **Plasma Shielding Designs**: If plasma shielding is part of the BIS design, simulations can explore different methods for generating and maintaining plasma, such as varying microwave frequencies or magnetic confinement techniques. The effectiveness of these methods in creating a stable and protective plasma barrier can be evaluated through these simulations.

Analyzing Simulation Data to Refine BIS Design

1. Evaluating Simulation Results

Analyzing the data generated from simulations is crucial for refining the BIS design. This process involves several steps:

- **Data Interpretation**: Simulation results provide insights into the behavior of the BIS under different conditions. For instance, the effectiveness of magnetic fields in deflecting bullets can be assessed by analyzing field strength, bullet trajectory changes, and any deviations from the intended path. Plasma behavior and its impact on bullet deflection are also analyzed for their effectiveness and stability.

- **Performance Metrics**: Key performance metrics, such as the degree of bullet deflection, energy consumption, and system stability, are extracted from the simulation data. These metrics help in evaluating the efficiency of different design configurations and identifying areas for improvement.

- **Sensitivity Analysis**: Sensitivity analysis is used to determine how variations in design parameters affect system performance. By adjusting parameters such as magnetic field strength, coil configurations, or plasma density, the simulation can reveal which factors are most critical to the BIS's effectiveness and which have minimal impact.

2. Refining the BIS Design

Based on the analysis of simulation data, the BIS design can be refined through iterative processes:

- **Design Optimization**: Optimization algorithms can be applied to find the best design parameters that maximize performance while minimizing drawbacks. Techniques such as genetic algorithms, simulated annealing, or gradient descent can be used to systematically explore the design space and identify optimal configurations.

- **Prototype Iteration**: Simulation results can guide the development of physical prototypes. By incorporating insights gained from simulations, prototypes can be built with improved designs and configurations. Iterative testing and refinement of prototypes based on further simulations help in achieving the desired performance.

- **Validation and Verification**: Simulations must be validated against experimental data to ensure their accuracy. Once physical prototypes are developed, testing them in real-world conditions allows for comparison with simulated results. Discrepancies between simulated and experimental data can reveal areas where the model may need adjustment or improvement.

3. Advanced Simulation Techniques

To enhance the accuracy and reliability of simulations, advanced techniques can be employed:

- **Multi-Physics Simulations**: Combining different physical phenomena, such as electromagnetism, fluid dynamics, and plasma physics, in a single simulation can provide a more comprehensive understanding of the BIS's behavior. Multi-physics simulations help in capturing complex interactions and improving the accuracy of predictions.

- **High-Performance Computing**: Large-scale simulations often require significant computational resources. Utilizing high-performance computing (HPC) clusters or cloud-based solutions can enable the simulation of complex models with higher resolution and accuracy.

- **Machine Learning Integration**: Machine learning algorithms can be integrated into the simulation process to enhance predictive capabilities and optimize design parameters. Techniques such as neural networks or reinforcement learning can analyze simulation data and provide insights for design improvements.

Conclusion

Simulating the Bullet Inversion System (BIS) is a crucial step in developing an effective and functional protective technology. By using advanced computational models and analyzing simulation data, designers can test various configurations, optimize performance, and refine the BIS design before physical prototypes are built. The iterative process of simulation, analysis, and refinement helps in achieving a more reliable and efficient BIS, paving the way for advancements in personal protection technology. Continued innovation in simulation techniques and computational methods will further enhance the development and optimization of BIS and similar technologies.

Prototype Development for Bullet Inversion System (BIS)

The development of prototypes for the Bullet Inversion System (BIS) is a critical phase in translating theoretical models and simulations into tangible, functional devices. This stage involves constructing small-scale experimental prototypes, rigorously testing them in controlled environments, and refining designs based on empirical data. Effective prototype development ensures that the BIS can be evaluated under realistic conditions and optimized for practical application.

Building Small-Scale Experimental BIS Prototypes

1. Design Considerations for Prototypes

Creating small-scale prototypes involves several key design considerations to ensure that the prototypes effectively test the core concepts of the BIS:

- **Component Selection**: The selection of components for the prototype is crucial. This includes choosing materials for magnetic field generators, power supplies, and any plasma generation systems. For instance, electromagnets may be constructed using soft ferromagnetic materials for the core and high-current wire for the coils. Similarly, for plasma generation, microwave generators and suitable ionization chambers must be selected.

- **Miniaturization**: Small-scale prototypes require careful miniaturization of components while maintaining functionality. This often involves using compact and high-performance components that can replicate the performance of full-scale systems on a reduced scale. Advanced fabrication techniques, such as precision machining or 3D printing, may be used to produce miniaturized components.

- **Integration**: Integrating various components into a cohesive prototype is essential. This includes ensuring that the magnetic field generators, power systems, and any plasma modules are effectively connected and interact as intended. Integration also involves designing and assembling the prototype's housing and support structures to accommodate all components and ensure stability.

2. Prototype Fabrication

The fabrication process for small-scale BIS prototypes includes several steps:

- **Blueprint and CAD Modeling**: Detailed blueprints and computer-aided design (CAD) models are created to guide the fabrication process. These models ensure that all components are accurately sized and positioned. CAD software can simulate the assembly and integration of parts, providing insights into potential issues before physical construction begins.

- **Component Manufacturing**: Components are manufactured according to the design specifications. This may involve machining metal parts, winding coils, or assembling electronic circuits. Precision is critical to ensure that the prototype operates as expected.

- **Assembly and Integration**: Once individual components are fabricated, they are assembled into the prototype. This includes mounting magnetic field generators, wiring power supplies, and setting up any plasma generation systems. Careful attention is paid to ensure that all connections are secure and that the prototype is properly aligned.

Testing Prototypes in Controlled Environments

1. Testing Procedures

Testing prototypes in controlled environments is essential for validating their performance and identifying areas for improvement:

- **Controlled Conditions**: Testing is conducted in environments where variables such as temperature, humidity, and electromagnetic interference can be precisely controlled. This ensures that the results are consistent and that the performance of the BIS prototype is accurately assessed.

- **Performance Evaluation**: Key performance metrics are evaluated during testing. For BIS prototypes, this includes measuring the strength and uniformity of magnetic fields, assessing the effectiveness of bullet deflection, and evaluating energy consumption and heat generation. Instrumentation such as magnetic field sensors, high-speed cameras, and thermocouples are used to collect data.

- **Safety Testing**: Safety is a critical aspect of prototype testing. Prototypes are tested to ensure they operate safely under all expected conditions. This includes verifying that cooling systems are effective, that electrical components are properly insulated, and that the device does not pose any risks to operators or bystanders.

2. Iterative Testing and Refinement

Testing is an iterative process that involves continuous refinement of the prototype based on test results:

- **Data Analysis**: Data collected during testing is analyzed to assess the performance of the prototype. This includes comparing the measured results to theoretical predictions and identifying any discrepancies. Data analysis helps in understanding the behavior of the BIS and in pinpointing areas where design improvements are needed.

- **Design Adjustments**: Based on the test results, design adjustments are made to improve the prototype's performance. This may involve modifying the configuration of magnetic field generators, adjusting power supply settings, or enhancing cooling systems. The adjusted design is then used to build a new prototype.

- **Re-testing**: The refined prototype is subjected to further testing to evaluate the impact of design changes. This iterative process continues until the prototype meets the desired performance criteria and operates reliably under various conditions.

3. Documentation and Analysis

Documenting the testing process and analyzing the results are crucial for refining the BIS design:

- **Test Reports**: Detailed reports are generated for each test, including data on performance metrics, safety evaluations, and observations. These reports provide a comprehensive record of the prototype's behavior and performance.

- **Design Evolution**: The evolution of the prototype design is tracked through documentation of each iteration. This helps in understanding how design changes impact performance and in making informed decisions for future improvements.

- **Feedback and Recommendations**: Feedback from testing is used to make recommendations for further development. This includes identifying areas where additional research or development may be required and suggesting improvements to enhance the BIS's functionality.

Conclusion

Prototype development is a critical phase in the creation of the Bullet Inversion System (BIS), involving the construction of small-scale experimental models and rigorous testing in controlled environments. By carefully designing and fabricating prototypes, and by conducting thorough testing and analysis, developers can validate theoretical models, optimize system performance, and address any issues before moving to full-scale production. The iterative process of prototype development ensures that the BIS can be effectively refined and improved, paving the way for a practical and functional protective technology.

Field Testing of Bullet Inversion System (BIS)

Field testing is a crucial phase in the development of the Bullet Inversion System (BIS), transitioning from theoretical models and controlled prototypes to real-world applications. This stage involves evaluating the BIS in actual environments to assess its effectiveness, reliability, and adaptability under various conditions. The results from field testing provide critical insights into the system's performance, guiding further refinement and ensuring that the BIS meets practical requirements.

Real-World Testing Scenarios for BIS Effectiveness

1. Controlled Environment Testing

Initially, field testing of the BIS is often conducted in controlled environments that simulate real-world conditions while allowing for precise monitoring and control. These environments are designed to test specific aspects of the BIS:

- **Testing Ranges**: Dedicated testing ranges are used to evaluate the BIS's performance in a controlled setting. These ranges can simulate various distances, angles, and velocities of incoming projectiles, allowing for comprehensive assessment of the system's deflection capabilities.

- **Simulated Threats**: In a controlled environment, various types of projectiles, including different calibers and materials, are used to test the BIS. Simulated threats help evaluate how well the system can handle different scenarios and ensure that it provides consistent protection across a range of potential threats.

- **Environmental Simulation**: Controlled environments can also simulate different environmental conditions, such as varying temperatures, humidity levels, and lighting conditions. This helps assess how environmental factors might impact the BIS's performance and reliability.

2. Real-World Deployment Scenarios

Field testing extends to real-world scenarios where the BIS is deployed in environments that closely resemble its intended application. These scenarios test the system's effectiveness in practical, operational conditions:

- **Urban Environments**: Testing in urban environments involves deploying the BIS in locations such as city streets, public spaces, or inside buildings. This scenario evaluates the system's performance in complex and dynamic settings, including factors such as obstacles, reflections, and varying levels of background electromagnetic interference.

- **High-Risk Situations**: Field testing also includes high-risk situations where the BIS is exposed to simulated threats in scenarios such as security checkpoints, military exercises, or law enforcement operations. These tests assess the system's ability to protect against actual threats and its functionality in high-pressure situations.

- **Wearable Configurations**: For BIS applications that involve wearable or portable devices, testing includes evaluating the system's performance when worn by individuals.

This scenario tests comfort, usability, and effectiveness in providing protection while the user is engaged in various activities.

Assessing Performance Under Various Conditions

1. Performance Metrics

Assessing the performance of the BIS during field testing involves evaluating several key metrics to determine how well the system meets its objectives:

- **Deflection Accuracy**: One of the primary metrics is the accuracy of bullet deflection. This involves measuring how effectively the BIS alters the trajectory of incoming projectiles and prevents impact with the protected individual. High-speed cameras and trajectory analysis tools are used to capture and analyze deflected projectiles.

- **System Reliability**: The reliability of the BIS is assessed by examining its performance over extended periods and under repeated use. This includes evaluating the durability of components, the stability of magnetic fields, and the consistency of plasma generation (if applicable). Reliability testing ensures that the BIS performs reliably in real-world conditions.

- **Energy Consumption**: Energy consumption is a critical metric, particularly for portable or wearable BIS devices. Testing measures the power required to operate the system effectively and evaluates the efficiency of energy storage and management. This helps identify potential improvements in power consumption and battery life.

- **User Experience**: For wearable BIS systems, user experience is assessed by gathering feedback from test participants regarding comfort, ease of use, and any impact on mobility. User experience testing ensures that the BIS is practical and functional for its intended users.

2. Adaptability and Robustness

Field testing evaluates the BIS's adaptability and robustness by exposing it to various real-world challenges:

- **Environmental Conditions**: The BIS is tested under different environmental conditions, including extreme temperatures, high humidity, and varying lighting conditions. This ensures that the system remains effective and operational in diverse settings.

- **Operational Stress**: Stress testing involves subjecting the BIS to high-stress scenarios, such as high-velocity projectiles or rapid field changes, to evaluate how well it maintains performance under pressure. This helps identify potential weaknesses and areas for improvement.

- **Interference and Obstructions**: The BIS is tested in environments with potential electromagnetic interference or physical obstructions to assess how these factors affect its performance. This includes evaluating how the system handles interference from other electronic devices or structures that may impact magnetic fields or plasma shielding.

3. Data Collection and Analysis

Comprehensive data collection and analysis are integral to field testing:

- **Instrumentation**: Various instruments are used to collect data during field tests, including sensors for magnetic field strength, cameras for trajectory analysis, and thermometers for temperature measurement. Accurate instrumentation ensures that performance metrics are reliably measured.

- **Data Analysis**: Data collected during field testing is analyzed to assess the BIS's effectiveness and identify areas for improvement. Statistical analysis and comparison with theoretical predictions help evaluate how well the BIS performs under real-world conditions and guide design adjustments.

- **Feedback Integration**: Feedback from test participants and observers is integrated into the analysis process. This feedback provides valuable insights into the practical aspects of the BIS, such as usability, comfort, and real-world effectiveness.

Conclusion

Field testing is a pivotal stage in the development of the Bullet Inversion System (BIS), providing essential insights into its real-world performance and effectiveness. By conducting tests in both controlled and actual environments, developers can assess the BIS's capabilities, reliability, and adaptability under various conditions. Evaluating key performance metrics, adapting to real-world challenges, and analyzing comprehensive data ensure that the BIS meets its design objectives and provides effective protection. The insights gained from field testing are instrumental in refining the BIS, optimizing its functionality, and advancing its potential applications in personal protection technology.

Safety Testing for the Bullet Inversion System (BIS)

Safety testing is an essential component of developing and deploying the Bullet Inversion System (BIS), ensuring that the system not only performs its intended function of bullet deflection but also operates safely for users and bystanders. This comprehensive evaluation process involves identifying potential risks, implementing mitigation strategies, and validating that all safety protocols are effectively addressed. The following discussion explores the key aspects of safety testing for the BIS, focusing on user and bystander safety, risk identification, and mitigation strategies.

Ensuring BIS Safety for Users and Bystanders

1. User Safety

The primary focus of safety testing is to protect the user of the BIS. Ensuring user safety involves evaluating how the system interacts with the user during operation and identifying any potential hazards associated with its use:

- **Operational Safety**: The BIS must be designed to operate safely in various environments and conditions. This includes evaluating the system's behavior under normal and extreme operational scenarios to ensure that it does not inadvertently cause harm. For instance, testing must confirm that magnetic fields do not interfere with other electronic devices or medical implants that users might have.

- **Electromagnetic Radiation**: The BIS, particularly if it involves strong magnetic fields or plasma generation, must be assessed for electromagnetic radiation emissions. Safety testing ensures that these emissions are within acceptable limits to prevent adverse effects on users and bystanders. This includes evaluating compliance with regulatory standards for electromagnetic radiation.

- **Ergonomics and Physical Interaction**: The design of the BIS should prioritize user comfort and ergonomics. Safety testing includes assessing the physical interaction between the BIS and the user to ensure that it does not cause discomfort or injury. For wearable devices, this involves evaluating the impact on mobility, posture, and overall ease of use.

- **Emergency Procedures**: Testing must include scenarios where the BIS may malfunction or fail. Emergency procedures should be developed and tested to ensure that users can safely disengage or deactivate the system if needed. This includes fail-safes and manual overrides to prevent potential accidents.

2. Bystander Safety

Ensuring the safety of bystanders is equally important, particularly in scenarios where the BIS is used in public or shared environments:

- **Shielding and Containment**: The BIS should be designed to prevent any unintended effects on people or objects outside its intended protection zone. Safety testing involves assessing the effectiveness of shielding and containment mechanisms to ensure that no hazardous effects, such as stray magnetic fields or plasma, impact bystanders.

- **Safety Zones**: Establishing safety zones around the BIS during operation is crucial. Testing includes evaluating the appropriate distance that bystanders should maintain to avoid potential risks. This involves determining safe operational ranges and ensuring that warnings or barriers are in place to keep people at a safe distance.

- **Emergency Response**: Safety testing must include evaluating the BIS's response to emergency situations involving bystanders. This includes assessing how the system reacts to sudden changes, malfunctions, or potential accidents and ensuring that it does not pose a risk to nearby individuals.

Identifying and Mitigating Potential Risks

1. Risk Identification

Identifying potential risks involves a thorough analysis of the BIS's design, functionality, and operating environment:

- **Failure Modes**: Assessing potential failure modes is a critical aspect of risk identification. This involves examining how various components of the BIS might fail and the potential consequences of such failures. For example, if an electromagnet fails, it could impact the system's ability to generate a magnetic field and potentially expose users to risk.

- **Interaction with Other Systems**: The BIS's interaction with other systems and devices must be evaluated to identify any risks associated with electromagnetic interference or other interactions. This includes ensuring compatibility with other technologies and assessing potential cross-system impacts.

- **Material and Component Safety**: Analyzing the safety of materials and components used in the BIS is essential. This includes evaluating the thermal properties of materials used in plasma generation and ensuring that they do not degrade or become hazardous under operational conditions.

2. Mitigation Strategies

Implementing effective mitigation strategies involves addressing identified risks and ensuring that the BIS operates safely:

- **Design Improvements**: Based on risk assessment findings, design improvements should be made to address potential hazards. This may involve incorporating additional shielding, enhancing cooling systems, or redesigning components to improve safety. For example, using advanced materials with better thermal and electromagnetic properties can enhance safety.

- **Safety Protocols**: Developing and implementing safety protocols is crucial for mitigating risks. This includes creating operational guidelines, safety checks, and maintenance procedures to ensure that the BIS remains safe throughout its lifecycle. Regular safety audits and inspections should be conducted to verify compliance with safety standards.

- **Training and Awareness**: Providing training and awareness programs for users is essential to ensure that they understand how to operate the BIS safely. This includes educating users about potential risks, emergency procedures, and safe practices. Training programs should be designed to address various scenarios and ensure that users are prepared for any situation.

- **Regulatory Compliance**: Ensuring that the BIS complies with relevant safety regulations and standards is crucial for mitigating risks. This includes adhering to guidelines set by regulatory bodies for electromagnetic fields, plasma generation, and general safety. Compliance with industry standards helps ensure that the BIS meets safety requirements and is suitable for deployment.

3. Continuous Improvement

Safety testing is an ongoing process that involves continuous improvement:

- **Feedback Integration**: Incorporating feedback from users, testers, and safety experts is essential for refining safety measures. Continuous feedback helps identify new risks and areas for improvement, leading to iterative design enhancements and updated safety protocols.

- **Technology Advancements**: Staying updated with advancements in technology and safety practices is crucial for maintaining the BIS's safety. Integrating new technologies and methods can enhance the effectiveness of safety measures and address emerging risks.

- **Post-Deployment Monitoring**: After deployment, monitoring the BIS in real-world conditions helps identify any unforeseen risks and assess the effectiveness of safety measures. Post-deployment monitoring includes gathering data on the system's performance, user experiences, and any incidents that occur, leading to further refinements and improvements.

Conclusion

Safety testing is a fundamental aspect of developing and deploying the Bullet Inversion System (BIS), ensuring that it operates safely for both users and bystanders. By identifying potential risks, implementing effective mitigation strategies, and continuously improving safety measures, developers can ensure that the BIS provides reliable protection while minimizing hazards. Comprehensive safety testing, including user and bystander safety evaluations, risk identification, and mitigation, is essential for delivering a robust and effective BIS that meets safety standards and performs reliably in real-world conditions.

Energy Efficiency Improvements for the Bullet Inversion System (BIS)

Energy efficiency is crucial in the development of advanced technologies such as the Bullet Inversion System (BIS). Enhancing energy efficiency not only reduces operational costs but also improves the overall feasibility and effectiveness of the BIS. This discussion explores strategies for enhancing energy efficiency in the BIS, as well as recent advances in energy-efficient technologies that can be leveraged to optimize the system's performance.

Strategies for Enhancing BIS Energy Efficiency

1. Optimizing Power Consumption

1.1. Efficient Power Management

Effective power management is essential for enhancing energy efficiency in the BIS. This involves implementing advanced power control systems that regulate the amount of energy used by various components of the BIS. Techniques such as dynamic power scaling and load balancing can help optimize energy consumption. For example, power management systems can adjust the energy supplied to magnetic field generators based on real-time operational demands, reducing waste and improving efficiency.

1.2. Minimizing Idle Power Usage

Reducing energy consumption during periods of inactivity is crucial. The BIS can incorporate sleep modes or low-power states that deactivate non-essential components when the system is not in active use. This approach involves designing the BIS to intelligently switch between active and standby modes, ensuring that power is only used when necessary. Implementing energy-efficient standby features can significantly reduce overall energy consumption.

2. Improving Component Efficiency

2.1. Advanced Magnetic Field Generators

Enhancing the efficiency of magnetic field generators is a key strategy for reducing energy consumption. This includes optimizing the design and materials used in electromagnets and permanent magnets. For instance, using high-performance superconducting materials can generate strong magnetic fields with minimal energy loss. Superconductors exhibit zero electrical resistance, which translates to highly efficient magnetic field generation.

2.2. Heat Management

Efficient heat management is vital for reducing energy waste. High-current electromagnets and plasma generators generate significant amounts of heat, which can lead to energy inefficiency if not properly managed. Implementing advanced cooling systems, such as heat sinks, liquid cooling, or phase-change materials, can effectively dissipate excess heat and maintain optimal operating temperatures. This prevents energy loss due to overheating and ensures consistent performance.

3. Integrating Energy Storage Solutions

3.1. High-Energy Batteries

High-energy batteries are essential for providing the necessary power to the BIS while maintaining energy efficiency. Advances in battery technology, such as lithium-sulfur or solid-state batteries, offer higher energy densities and longer lifespans compared to traditional battery types. Utilizing these advanced batteries can reduce the frequency of recharging and improve overall system efficiency.

3.2. Supercapacitors

Supercapacitors are another energy storage solution that can enhance energy efficiency. They are capable of storing and delivering large amounts of energy quickly, making them suitable for applications requiring rapid bursts of power. By integrating supercapacitors with traditional batteries, the BIS can benefit from both high energy density and fast power delivery, optimizing overall efficiency.

4. Efficient Energy Conversion

4.1. Advanced Power Electronics

Utilizing advanced power electronics can improve energy conversion efficiency in the BIS. High-efficiency converters and inverters reduce energy losses during the conversion of electrical energy from one form to another. For instance, switching from traditional silicon-based power devices to wide-bandgap materials like silicon carbide (SiC) or gallium nitride (GaN) can enhance power conversion efficiency and reduce energy losses.

4.2. Regenerative Systems

Incorporating regenerative systems can further improve energy efficiency. Regenerative braking, for example, captures and reuses energy during braking or deceleration phases. While not directly applicable to all aspects of the BIS, regenerative systems can be adapted to recover energy from other processes or components, such as cooling systems, and redirect it for use elsewhere in the system.

Advances in Energy-Efficient Technologies

1. High-Performance Superconductors

Recent advancements in superconducting materials have significantly improved energy efficiency. High-temperature superconductors (HTS), such as yttrium barium copper oxide (YBCO), operate at relatively higher temperatures compared to traditional low-temperature superconductors (LTS). These materials can generate stronger magnetic fields with minimal energy loss, making them ideal for high-performance applications in the BIS. Ongoing research into new superconducting materials aims to further enhance their performance and reduce cooling requirements.

2. Energy-Efficient Cooling Technologies

Advances in cooling technologies have also contributed to energy efficiency improvements. Techniques such as microchannel cooling, which uses small-scale channels to dissipate heat more effectively, can enhance the performance of heat-sensitive components in the BIS. Additionally, phase-change materials (PCMs) offer efficient thermal management by absorbing and releasing heat during phase transitions. These technologies help maintain optimal operating temperatures while minimizing energy consumption.

3. Advanced Battery Technologies

Innovations in battery technologies are driving improvements in energy efficiency. Solid-state batteries, for example, offer higher energy densities and improved safety compared to traditional lithium-ion batteries. They also have lower self-discharge rates, which enhances overall efficiency. Research into alternative chemistries, such as lithium-sulfur or sodium-ion batteries, is exploring ways to further increase energy storage capabilities and reduce costs.

4. Smart Power Management Systems

The development of smart power management systems is another significant advancement. These systems use artificial intelligence (AI) and machine learning algorithms to optimize energy usage based on real-time data and predictive analytics. By analyzing operational patterns and environmental conditions, smart power management systems can adjust energy consumption dynamically, improving overall efficiency and reducing waste.

5. Energy Harvesting Technologies

Energy harvesting technologies offer opportunities for enhancing energy efficiency by capturing and utilizing ambient energy sources. For example, piezoelectric materials can convert mechanical vibrations into electrical energy, which can be used to power low-energy components of the BIS. Similarly, thermoelectric materials can convert temperature gradients into electrical energy. Integrating energy harvesting technologies can supplement traditional power sources and improve overall efficiency.

6. Miniaturization and Integration

Advancements in miniaturization and integration technologies enable the development of more compact and energy-efficient systems. Techniques such as integrated circuit design and additive manufacturing allow for the creation of smaller, more efficient components. By reducing the size and complexity of the BIS, these technologies can minimize energy consumption and improve overall system performance.

Conclusion

Enhancing energy efficiency is a critical aspect of developing the Bullet Inversion System (BIS) and other advanced technologies. By optimizing power consumption, improving component efficiency, integrating advanced energy storage solutions, and leveraging recent technological advances, developers can achieve significant improvements in energy efficiency. Continued research and innovation in energy-efficient technologies will drive the future of the BIS, enabling more effective and sustainable personal protection systems.

Material Durability and Performance in High-Energy Environments

The development of materials capable of enduring high-energy environments is pivotal for advancing technologies such as the Bullet Inversion System (BIS). High-energy environments impose extreme conditions, including intense electromagnetic fields, high temperatures, and rapid mechanical stresses. To ensure reliability and effectiveness, materials used in such systems must exhibit exceptional durability and performance. This discussion explores the theoretical, practical, and logical aspects of developing durable materials, as well as methods for testing material degradation over time.

Developing Materials for High-Energy Environments

1. Theoretical Foundations

1.1. Material Properties and Energy Interactions

The theoretical foundation for developing materials suitable for high-energy environments involves understanding how different properties interact with energy inputs. Key properties include thermal conductivity, electrical conductivity, magnetic permeability, and mechanical strength. For instance, materials with high thermal conductivity can dissipate heat more effectively, while those with high electrical conductivity can efficiently manage electrical currents.

1.2. Advanced Material Science

Recent advancements in material science have led to the development of materials with tailored properties for specific high-energy applications. Nanotechnology and material engineering allow for the creation of composites and hybrid materials that combine desirable attributes. For example, incorporating nanoparticles into polymers can enhance their thermal stability and mechanical strength, making them more suitable for high-energy environments.

2. Practical Considerations

2.1. Selection of High-Performance Materials

Choosing the right materials for high-energy environments involves selecting substances that can withstand extreme conditions without degrading. Key considerations include:

- **Thermal Stability:** Materials must resist thermal degradation and maintain their properties at elevated temperatures. Materials such as ceramics, refractory metals, and high-temperature polymers are often used for their superior thermal stability.

- **Mechanical Strength:** High-energy environments can induce mechanical stresses, so materials need high tensile strength and resistance to fatigue. Advanced composites and reinforced alloys are examples of materials engineered for enhanced mechanical performance.

- **Electromagnetic Compatibility:** In applications involving strong electromagnetic fields, materials must exhibit suitable electromagnetic properties. Superconductors and magnetic shielding materials are examples of substances engineered to handle intense electromagnetic environments.

2.2. Material Processing Techniques

Advanced processing techniques can improve material performance by enhancing their structural integrity and resistance to high-energy conditions. Techniques such as sintering, alloying, and additive manufacturing enable the production of materials with optimized properties. For example, sintering can enhance the density and mechanical properties of ceramics, while additive manufacturing allows for precise control over material composition and structure.

Testing Material Degradation Over Time

1. Theoretical Aspects

1.1. Degradation Mechanisms

Understanding material degradation mechanisms is crucial for predicting the longevity and reliability of materials in high-energy environments. Key degradation mechanisms include:

- **Thermal Degradation:** Exposure to high temperatures can cause chemical reactions or phase changes that alter material properties. For example, polymers may undergo thermal oxidative degradation, leading to reduced strength and flexibility.

- **Mechanical Wear and Fatigue:** Repeated mechanical stresses can lead to wear and fatigue, resulting in cracks or structural failures. Testing must account for cyclic loading conditions to assess long-term durability.

- **Electromagnetic Effects:** Materials exposed to strong electromagnetic fields can experience changes in their electromagnetic properties, such as magnetization or electrical resistance. Understanding these effects is essential for ensuring material stability.

1.2. Modeling and Simulation

Theoretical modeling and simulation techniques are used to predict material behavior under various conditions. Computational models can simulate the effects of thermal, mechanical, and electromagnetic stresses on material performance, allowing for the prediction of degradation patterns and the evaluation of material longevity.

2. Practical Testing Methods

2.1. Accelerated Aging Tests

Accelerated aging tests simulate long-term exposure to high-energy conditions in a shorter time frame. These tests involve subjecting materials to extreme temperatures, high radiation levels, or intense mechanical stresses to accelerate degradation processes. Results from these tests provide insights into material longevity and performance under real-world conditions.

2.2. Mechanical Testing

Mechanical testing assesses a material's ability to withstand physical stresses and strains over time. Key tests include:

- **Tensile Testing:** Measures the material's strength and ductility by applying a tensile load until failure. This test helps evaluate how materials respond to mechanical stresses.

- **Fatigue Testing:** Assesses the material's resistance to cyclic loading by applying repeated stresses until failure occurs. This test provides information on the material's durability under dynamic conditions.

- **Hardness Testing:** Determines the material's resistance to indentation or abrasion. Hardness testing is useful for evaluating wear resistance.

2.3. Thermal Testing

Thermal testing evaluates how materials perform under varying temperature conditions. Key tests include:

- **Thermogravimetric Analysis (TGA):** Measures changes in material mass as a function of temperature. This test provides information on thermal stability and decomposition temperatures.
- **Differential Scanning Calorimetry (DSC):** Measures heat flow associated with phase transitions or chemical reactions. DSC provides insights into thermal properties such as melting points and glass transition temperatures.

2.4. Electromagnetic Testing

For materials exposed to electromagnetic fields, testing involves assessing changes in electromagnetic properties over time. Key tests include:

- **Magnetic Susceptibility Testing:** Measures the material's response to external magnetic fields, providing information on changes in magnetic properties.

- **Electrical Conductivity Testing:** Assesses changes in electrical conductivity due to exposure to electromagnetic fields. This test is important for materials used in electromagnetic shielding or high-current applications.

Conclusion

The development of materials capable of withstanding high-energy environments is critical for advancing technologies such as the Bullet Inversion System (BIS). By understanding theoretical principles, considering practical material selection and processing techniques, and implementing rigorous testing methods, researchers and engineers can ensure the durability and performance of materials in extreme conditions. Continued advancements in material science and testing methodologies will drive the development of more reliable and effective high-energy systems, contributing to the success of innovative technologies.

Future Material Research: Innovations in Magnetic Materials and Alternatives for Bullet Influence

The field of material science is on the brink of transformative advancements that hold the potential to revolutionize various technologies, including those involved in high-energy and defensive applications such as the Bullet Inversion System (BIS). Future material research is poised to uncover new magnetic materials with superior properties and explore alternative materials that could influence bullets in innovative ways. This discussion delves into the theoretical, practical, logical, and expansive aspects of these exciting research avenues.

Investigating New Magnetic Materials with Superior Properties

1. Theoretical Foundations

1.1. Advances in Magnetic Materials

The theoretical exploration of new magnetic materials involves understanding and manipulating their magnetic properties to achieve enhanced performance. Magnetic materials are categorized based on their magnetic behaviors, such as ferromagnetism, antiferromagnetism, ferrimagnetism, and paramagnetism. Research into new materials often focuses on:

- **High Magnetic Permeability:** Materials with high magnetic permeability can concentrate magnetic flux lines, resulting in stronger and more efficient magnetic fields. Novel materials that exhibit high permeability and low core losses are sought for applications in electromagnetic devices.
- **High Magnetic Saturation:** Magnetic saturation refers to the maximum magnetization a material can achieve under an applied magnetic field. Materials with high saturation magnetization can generate stronger magnetic fields without saturating, making them ideal for high-performance applications.

1.2. Emerging Material Systems

Emerging materials with unique magnetic properties are being explored for their potential applications. Examples include:

- **Metamaterials:** Engineered composites with designed electromagnetic properties can exhibit unusual responses to magnetic fields, such as negative permeability or tailored field distributions. These materials offer possibilities for creating advanced magnetic field configurations.
- **Topological Insulators:** These materials possess unique electronic surface states that can influence magnetic behavior. Research into topological insulators explores their potential for novel magnetic applications and interactions.

2. Practical Research Approaches

2.1. Synthesis and Fabrication

Developing new magnetic materials involves synthesizing and fabricating them with precise control over their composition and structure. Techniques include:

- **Chemical Vapor Deposition (CVD):** CVD allows for the deposition of thin magnetic films with controlled thickness and uniformity. This technique is used to create high-quality magnetic layers for various applications.
- **Sol-Gel Processes:** Sol-gel methods are used to prepare magnetic ceramics and composites by transitioning from a sol (liquid) to a gel (solid) state. This approach enables the production of materials with tailored magnetic properties.

2.2. Characterization and Testing

Once synthesized, new magnetic materials undergo rigorous characterization to assess their properties. Key techniques include:

- **Magnetometry:** Instruments such as vibrating sample magnetometers (VSM) and superconducting quantum interference devices (SQUID) measure magnetic properties like saturation magnetization and coercivity.
- **X-ray Diffraction (XRD):** XRD provides information on the crystal structure of magnetic materials, which influences their magnetic behavior. Analyzing the crystallographic phases helps in optimizing material performance.

Exploring Alternative Materials for Bullet Influence

1. Theoretical Foundations

1.1. Non-Magnetic Methods for Bullet Deflection

Exploring alternative materials for influencing bullets involves investigating non-magnetic methods that can alter a bullet's trajectory or impact. Key areas include:

- **Electromagnetic Induction:** By creating electromagnetic fields that induce currents in conductive bullets, it may be possible to alter their paths. Materials that can generate strong, controlled electromagnetic fields are essential for this approach.
- **Plasma Shields:** Generating a plasma field around the target could deflect or neutralize bullets. Plasma, being a collection of ionized gases, interacts with bullets in a way that could potentially influence their trajectories.

1.2. Advanced Material Systems

Research into advanced material systems explores the use of materials with unconventional properties to influence bullets:

- **Smart Materials:** Materials that change properties in response to external stimuli, such as magnetic or electric fields, could be used to develop adaptive systems that alter their behavior in the presence of bullets.
- **High-Energy Density Materials:** Materials capable of absorbing or dissipating large amounts of energy can be explored to create barriers that effectively neutralize bullets upon impact.

2. Practical Research Approaches

2.1. Material Development and Integration

Developing alternative materials for bullet influence involves integrating them into systems designed to interact with projectiles. Practical approaches include:

- **Composite Materials:** Combining materials with different properties, such as conductive and insulating layers, can create composites that influence bullets through electromagnetic or energy absorption mechanisms.
- **Material Coatings:** Applying advanced coatings to surfaces can modify their interaction with bullets. For example, coatings that react to impact or change properties under stress could offer new methods for bullet influence.

2.2. Experimental Validation

Practical experimentation is crucial for validating the effectiveness of alternative materials. Key methods include:

- **Ballistics Testing:** Conducting controlled ballistics tests to assess how materials affect bullet trajectories or impact behavior. This includes firing projectiles at materials and measuring deflection, penetration, or energy absorption.
- **Field Testing:** Real-world testing scenarios provide insights into how alternative materials perform under actual conditions. This involves deploying materials in practical environments and evaluating their effectiveness in dynamic situations.

3. Future Directions and Integration

3.1. Cross-Disciplinary Research

The future of material research in high-energy applications will increasingly involve cross-disciplinary approaches. Integrating insights from physics, materials science, and engineering can lead to innovative solutions. Collaboration between researchers in different fields will accelerate the development of advanced materials with superior properties.

3.2. Technological Integration

Advancements in materials science will be complemented by improvements in related technologies. For example, integrating advanced magnetic materials with high-efficiency electromagnetic systems can enhance performance and enable new applications. Similarly, combining smart materials with sensor technologies can create adaptive systems that respond to external threats more effectively.

Conclusion

Future material research holds significant promise for advancing technologies such as the Bullet Inversion System (BIS). Investigating new magnetic materials with superior properties and exploring alternative materials for bullet influence are key areas of focus. By leveraging theoretical insights, practical development approaches, and experimental validation, researchers can develop materials that push the boundaries of current technologies. The continued evolution of material science, coupled with interdisciplinary collaboration, will drive innovation and lead to groundbreaking advancements in high-energy and defensive applications.

Advances in Energy Storage: Innovations in Battery Technology and Supercapacitors

Energy storage is a critical aspect of modern technology, particularly in advanced systems like the Bullet Inversion System (BIS), which demands high efficiency and rapid energy deployment. As technology evolves, innovations in energy storage solutions, including batteries and supercapacitors, are pivotal for enhancing system performance and reliability. This discussion explores the theoretical, practical, logical, and expansive aspects of these advancements.

Innovations in Battery Technology for BIS

1. Theoretical Foundations

1.1. Energy Density and Efficiency

Energy density is a crucial parameter for batteries used in high-performance applications. It defines the amount of energy stored per unit volume or weight. Advanced battery technologies aim to maximize energy density while improving efficiency. Key theoretical concepts include:

- **Specific Energy:** This is the amount of energy a battery can store relative to its mass, often expressed in watt-hours per kilogram (Wh/kg). Innovations in battery chemistry, such as lithium-sulfur or lithium-air, seek to increase specific energy compared to traditional lithium-ion batteries.
- **Energy Efficiency:** This represents the ratio of energy output to energy input. High-efficiency batteries minimize energy losses during charging and discharging processes, enhancing overall system performance. Techniques such as advanced electrode materials and improved electrolyte formulations contribute to better energy efficiency.

1.2. Advanced Battery Chemistries

New battery chemistries are being explored to enhance energy storage capabilities:

- **Lithium-Sulfur Batteries:** These batteries offer a high theoretical energy density due to the high energy capacity of sulfur. They also benefit from lower costs compared to lithium-ion technologies. However, challenges such as sulfur's poor conductivity and volume expansion need to be addressed.

- **Solid-State Batteries:** Utilizing a solid electrolyte instead of a liquid or gel, solid-state batteries provide higher energy density, improved safety, and longer cycle life. Research focuses on developing stable solid electrolytes and efficient solid-electrolyte interfaces.

- **Lithium-Air Batteries:** These batteries have the potential for extremely high energy densities due to the use of oxygen from the air as a reactant. Practical challenges include the management of air flow and the stability of battery components.

2. Practical Applications

2.1. Enhanced Battery Designs

Practical advancements in battery technology involve designing and manufacturing batteries with improved performance characteristics:

- **High-Capacity Electrodes:** Innovations in electrode materials, such as graphene or silicon-based anodes, enhance the capacity and efficiency of batteries. These materials offer higher charge storage compared to traditional graphite electrodes.
- **Advanced Electrolytes:** New electrolyte formulations, including solid and gel electrolytes, aim to improve battery safety and performance. Research into non-flammable and high-conductivity electrolytes addresses key safety and efficiency concerns.

2.2. Manufacturing and Integration

Efficient manufacturing processes and integration techniques are essential for deploying advanced batteries in BIS applications:

- **Precision Manufacturing:** Advanced manufacturing techniques, such as roll-to-roll processing for flexible batteries, enable the production of high-quality and cost-effective battery components.
- **Battery Management Systems (BMS):** Integrated BMS technologies monitor and manage battery performance, ensuring optimal operation and safety. Features include charge balancing, temperature monitoring, and fault detection.

Exploring Supercapacitors and Other Energy Storage Solutions

1. Theoretical Foundations

1.1. Capacitance and Power Density

Supercapacitors, also known as ultracapacitors, are energy storage devices with high power density and rapid charge-discharge capabilities. Key theoretical aspects include:

- **Capacitance:** This defines a supercapacitor's ability to store charge, measured in farads (F). Supercapacitors achieve high capacitance through the use of high-surface-area electrodes and porous materials.
- **Power Density:** Supercapacitors excel in power density, which is the rate at which energy can be delivered or absorbed. This is crucial for applications requiring rapid bursts of power, such as in BIS systems where quick responses to threats are necessary.

1.2. Types of Supercapacitors

Various supercapacitor technologies are explored for their potential benefits and applications:

- **Electric Double-Layer Capacitors (EDLCs):** These supercapacitors store energy through the electrostatic separation of charges at the electrode-electrolyte interface. They offer high power density and long cycle life but have limited energy density.

- **Pseudocapacitors:** These supercapacitors utilize redox reactions at the electrode surface to store energy, providing higher energy density compared to EDLCs. They combine characteristics of both capacitors and batteries.

- **Hybrid Supercapacitors:** Combining features of supercapacitors and batteries, hybrid supercapacitors aim to enhance both energy and power density. They integrate battery-like charge storage mechanisms with supercapacitor technology.

2. Practical Applications

2.1. Integration into BIS Systems

Integrating supercapacitors into BIS systems requires addressing practical considerations for optimal performance:

- **Energy Management:** Supercapacitors are often used in conjunction with batteries to provide rapid bursts of power while batteries handle sustained energy needs. Energy management systems coordinate the operation of both components for efficient energy utilization.
- **Thermal Management:** Supercapacitors can experience significant heat generation during high-power operations. Effective thermal management systems, including cooling solutions and heat dissipation techniques, ensure reliable performance and longevity.

2.2. Advanced Supercapacitor Designs

Research into advanced supercapacitor designs focuses on improving their performance and integration:

- **Nanomaterials:** The use of nanomaterials, such as carbon nanotubes or graphene, enhances the surface area and conductivity of supercapacitors. These materials contribute to higher capacitance and better power density.
- **Flexible and Lightweight Designs:** Developing flexible and lightweight supercapacitors is crucial for applications requiring conformable and portable energy storage solutions. Innovations in materials and manufacturing techniques enable the production of flexible supercapacitor components.

3. Future Directions

3.1. Hybrid Energy Storage Systems

Combining batteries and supercapacitors into hybrid energy storage systems offers a balanced approach to energy and power needs. Future research will focus on optimizing the integration of these technologies to maximize their combined benefits.

3.2. Advances in Energy Storage Materials

Research into new materials for both batteries and supercapacitors will continue to drive advancements in energy storage. Innovations in material science, including the development of high-capacity electrodes and advanced electrolytes, will enhance performance and efficiency.

3.3. Sustainable and Scalable Solutions

Future energy storage solutions will prioritize sustainability and scalability. Research into recyclable materials, environmentally friendly manufacturing processes, and scalable production techniques will support the development of energy storage technologies that align with global sustainability goals.

Conclusion

Advances in energy storage technologies, including innovations in battery technology and supercapacitors, are crucial for enhancing the performance and efficiency of systems like the

Bullet Inversion System (BIS). By exploring new battery chemistries, optimizing supercapacitor designs, and integrating these technologies, researchers and engineers are paving the way for more effective and reliable energy storage solutions. Continued research and development will drive further innovations, ensuring that energy storage technologies meet the demands of high-performance applications and contribute to the advancement of modern technology.

Integration with Protective Technologies: Combining BIS with Advanced Body Armor and Automated Threat Detection Systems

The evolution of personal protective technologies has reached an inflection point where integration and synergy between different systems can significantly enhance overall safety and effectiveness. The Bullet Inversion System (BIS), a conceptual technology designed to deflect or neutralize incoming bullets using magnetic and electromagnetic fields, represents a groundbreaking advancement in personal protection. To maximize its potential, BIS must be seamlessly integrated with advanced body armor and automated threat detection systems. This discussion explores the theoretical, practical, logical, and expansive aspects of this integration.

Combining BIS with Advanced Body Armor

1. Theoretical Foundations

1.1. Synergy Between BIS and Body Armor

The BIS's primary function—deflecting or neutralizing bullets—can be complemented by traditional body armor, which provides physical protection against ballistic threats. Theoretical integration involves understanding how these technologies can work in tandem:

- **Complementary Protection:** While BIS aims to alter the trajectory or disrupt the functioning of bullets, body armor serves as a physical barrier to absorb and distribute the impact energy. Combining these technologies could offer multi-layered protection, where BIS handles the initial deflection and body armor provides additional resistance against residual impacts.
- **Dynamic Response:** BIS operates in real-time, responding to incoming threats by generating magnetic fields or plasma shields. Advanced body armor can be designed to adapt to the changes in the threat environment, such as incorporating materials that harden or change properties in response to electromagnetic fields generated by BIS.

1.2. Material Interactions

The interaction between BIS-generated magnetic fields and body armor materials requires detailed theoretical analysis:

- **Electromagnetic Interference:** The magnetic fields produced by BIS may affect electronic components or sensors embedded in body armor. Understanding these interactions is crucial for designing armor that can function effectively without interference.

- **Material Compatibility:** Body armor materials, such as Kevlar or advanced ceramics, must be tested for compatibility with BIS technologies. Research into materials that can withstand both ballistic impacts and electromagnetic fields is essential for successful integration.

2. Practical Applications

2.1. Integrated Design Approaches

Designing a combined BIS and body armor system involves practical considerations to ensure functionality and comfort:

- **Layered Systems:** One approach is to create a layered protective system where BIS components are integrated into the outer layer, while body armor constitutes the inner layers. This configuration allows the BIS to handle initial threats and the body armor to provide additional protection.
- **Modular Designs:** Modular designs enable the BIS and body armor to be used independently or together, depending on the threat level. This flexibility allows users to adapt their protective gear based on specific needs and scenarios.

2.2. Testing and Validation

Comprehensive testing is required to validate the effectiveness and safety of integrated systems:

- **Field Trials:** Conducting field trials with integrated BIS and body armor systems helps assess their performance under real-world conditions. These trials should simulate various threat scenarios to evaluate the combined system's effectiveness.
- **User Comfort:** Ensuring that the integrated system is comfortable and does not impede movement is crucial. Ergonomic design considerations, such as weight distribution and flexibility, are essential for user comfort and operational effectiveness.

Integrating Automated Threat Detection Systems

1. Theoretical Foundations

1.1. Real-Time Threat Detection

Automated threat detection systems use advanced sensors and algorithms to identify and assess potential threats in real-time. Theoretical integration with BIS involves:

- **Sensor Fusion:** Combining data from multiple sensors, such as radar, lidar, and infrared, to provide a comprehensive threat assessment. Integrating this data with BIS allows for a more accurate and timely response to incoming threats.

- **Predictive Analytics:** Automated systems can use predictive analytics to anticipate threat trajectories and adjust BIS parameters accordingly. This proactive approach enhances the system's ability to neutralize threats before they reach the user.

1.2. Communication Systems

Effective integration requires robust communication systems between BIS and automated threat detection technologies:

- **Data Transmission:** High-speed data transmission is necessary for real-time communication between sensors, threat detection systems, and BIS. Ensuring low latency and high reliability is crucial for effective threat neutralization.
- **Algorithmic Coordination:** Algorithms that process sensor data and control BIS responses must be synchronized. This coordination ensures that BIS can respond accurately and promptly to detected threats.

2. Practical Applications

2.1. Integrated System Design

Designing an integrated BIS and automated threat detection system involves several practical considerations:

- **Sensor Integration:** Sensors must be embedded into the BIS framework or worn as part of the protective gear. These sensors collect data on potential threats and communicate with the BIS to adjust its response in real-time.
- **User Interface:** Developing an intuitive user interface allows operators to monitor threat information and control BIS settings. The interface should provide clear, actionable insights and allow for manual overrides if necessary.

2.2. Testing and Optimization

Optimizing the integrated system requires rigorous testing and refinement:

- **Simulation Testing:** Computer simulations can model various threat scenarios and evaluate the performance of the integrated BIS and automated threat detection system. This testing helps identify potential issues and areas for improvement.
- **Field Testing:** Field testing with actual threat scenarios provides valuable insights into the system's performance and reliability. This testing helps fine-tune the integration and ensures that the system meets real-world operational requirements.

Expansive Considerations

1. Future Developments

Future advancements in protective technologies will likely focus on enhancing the integration of BIS with body armor and automated threat detection systems:

- **Advanced Materials:** Research into new materials that can interact seamlessly with BIS technologies and provide superior protection will drive future developments. These materials could offer improved performance and functionality in various threat scenarios.
- **AI and Machine Learning:** Incorporating AI and machine learning into threat detection and BIS control systems will enhance their ability to predict and respond to threats. These technologies can improve accuracy, reduce false positives, and adapt to evolving threat landscapes.

2. Interdisciplinary Collaboration

Successful integration of BIS with advanced body armor and automated threat detection systems requires interdisciplinary collaboration:

- **Engineering and Materials Science:** Collaboration between engineers and materials scientists is essential for developing effective and compatible protective technologies. This partnership can lead to innovations in materials and design.
- **Data Science and AI:** Data scientists and AI experts play a crucial role in developing algorithms for threat detection and system coordination. Their expertise ensures that integrated systems can process and respond to data efficiently.

Conclusion

The integration of the Bullet Inversion System (BIS) with advanced body armor and automated threat detection systems represents a significant leap in personal protection technology. By combining these technologies, it is possible to create a multi-layered, adaptive defense system that offers enhanced protection against a wide range of threats. Theoretical understanding, practical design considerations, and expansive research are essential for achieving effective integration and advancing the field of personal protection. As technology continues to evolve, the integration of BIS with other protective technologies will play a crucial role in shaping the future of safety and security.

Legal and Ethical Considerations: Addressing BIS Use and Ethical Implications of Personal Defense Technology

As the development of innovative personal defense technologies, such as the Bullet Inversion System (BIS), progresses, it is imperative to navigate the complex legal and ethical landscape surrounding their use. BIS, which aims to neutralize or deflect incoming bullets using magnetic fields, represents a significant advancement in personal protection. However, its implementation and deployment involve intricate legal and ethical considerations that must be addressed to ensure responsible and lawful use.

Addressing Legal Issues Related to BIS Use

1. Regulatory Compliance

1.1. Firearms and Ammunition Laws

The introduction of BIS technology raises questions about its compatibility with existing firearms and ammunition regulations. The legal framework governing firearms often includes stipulations about modifications or enhancements that could affect their operation. Regulatory compliance involves:

- **Modification Restrictions:** Many jurisdictions have strict laws regulating modifications to firearms, including the introduction of technologies that could alter their intended functionality. BIS must be assessed for compliance with these laws to avoid potential legal conflicts.
- **Ammunition Specifications:** Since BIS affects the trajectory of bullets, it is necessary to consider how this technology interacts with various types of ammunition. Legal issues could arise if BIS causes unintended consequences, such as increased risk or unintended harm.

1.2. Public Safety Regulations

Public safety regulations address the potential risks associated with advanced defense technologies. BIS integration must adhere to these regulations:

- **Safety Standards:** Compliance with safety standards ensures that BIS does not pose a risk to users or bystanders. These standards may include requirements for rigorous testing, certification, and adherence to safety protocols.
- **Liability and Insurance:** Legal liability for any harm caused by BIS, whether intentional or accidental, must be clearly defined. This includes determining insurance requirements and coverage for potential incidents involving BIS.

2. Privacy and Surveillance

2.1. Data Collection and Privacy

BIS, particularly when integrated with automated threat detection systems, may involve data collection related to threats and user behavior. Legal considerations include:

- **Data Protection Laws:** Compliance with data protection laws, such as the General Data Protection Regulation (GDPR) or the California Consumer Privacy Act (CCPA), is essential. These laws govern the collection, storage, and use of personal data, including data related to BIS operations.
- **User Consent:** Obtaining informed consent from users regarding the collection and use of their data is crucial. Clear communication about data handling practices and user rights is necessary to comply with privacy regulations.

2.2. Surveillance Concerns

The integration of BIS with surveillance and threat detection systems may raise concerns about privacy and surveillance:

- **Surveillance Legislation:** Laws governing surveillance, such as those related to wiretapping or video monitoring, must be considered. The deployment of BIS should not infringe upon privacy rights or exceed legal surveillance limits.
- **Ethical Use of Surveillance Data:** Ethical considerations include ensuring that surveillance data is used responsibly and not for purposes beyond its intended scope. This involves establishing guidelines for data access, usage, and protection.

Ethical Implications of Personal Defense Technology

1. Personal Autonomy and Responsibility

1.1. User Empowerment

The deployment of advanced personal defense technologies, like BIS, raises ethical questions about the balance between user empowerment and responsibility:

- **Autonomy:** Users of BIS should be empowered to make informed decisions about their personal safety. However, this empowerment must be balanced with an understanding of the potential risks and limitations of the technology.
- **Responsibility:** Users must be educated about the ethical implications of using BIS and the potential consequences of its deployment. This includes understanding the technology's limitations and the responsibility associated with its use.

1.2. Ethical Decision-Making

Ethical decision-making involves assessing the impact of BIS on individual and societal levels:

- **Use in Self-Defense:** BIS's primary purpose is to enhance personal safety. Ethical considerations include ensuring that its use is confined to legitimate self-defense scenarios and does not lead to unnecessary harm or escalation.
- **Potential for Misuse:** Addressing the potential for misuse of BIS, such as using it for aggressive or unlawful purposes, is crucial. Developing safeguards and guidelines to prevent misuse is an ethical imperative.

2. Societal Impact

2.1. Equity and Access

The availability and affordability of BIS technology may raise ethical concerns related to equity and access:

- **Access to Technology:** Ensuring that BIS technology is accessible to all individuals, regardless of socioeconomic status, is an ethical consideration. Disparities in access could exacerbate existing inequalities in personal safety.
- **Regulation of Distribution:** Ethical distribution practices must be established to prevent the technology from falling into the wrong hands or being used inappropriately.

2.2. Public Perception and Trust

The introduction of advanced defense technologies can influence public perception and trust:

- **Transparency:** Providing transparency about the capabilities, limitations, and risks associated with BIS is essential for building public trust. Clear communication and educational efforts can help mitigate concerns and foster acceptance.
- **Ethical Marketing:** Marketing practices for BIS should avoid exaggeration or misrepresentation of its capabilities. Ethical marketing ensures that potential users have realistic expectations and understand the technology's role in personal safety.

3. Ethical Governance and Regulation

3.1. Developing Ethical Frameworks

Establishing ethical frameworks and guidelines for BIS technology is vital for ensuring responsible use:

- **Ethical Committees:** Forming ethical committees or advisory boards to provide guidance on BIS development and deployment can help address ethical dilemmas and ensure adherence to ethical standards.
- **Regulatory Bodies:** Collaboration with regulatory bodies to develop comprehensive guidelines and standards for BIS technology is essential. These guidelines should address legal, ethical, and safety considerations.

3.2. Continuous Review and Adaptation

Ethical considerations for BIS technology must be reviewed and adapted as the technology evolves:

- **Ongoing Evaluation:** Regular evaluation of BIS technology's impact on society and individual rights is necessary to address emerging ethical issues and adapt guidelines accordingly.
- **Stakeholder Engagement:** Engaging with stakeholders, including users, legal experts, ethicists, and policymakers, ensures that ethical considerations are comprehensively addressed and integrated into the technology's development and deployment.

Conclusion

The integration of the Bullet Inversion System (BIS) with advanced protective technologies presents both legal and ethical challenges that must be carefully managed. Addressing legal issues related to BIS use involves navigating regulatory compliance, privacy concerns, and liability considerations. Ethical implications encompass personal autonomy, societal impact, and the responsible use of technology. By developing robust legal frameworks, ethical guidelines, and transparent practices, stakeholders can ensure that BIS and similar technologies are used in ways that enhance personal safety while respecting individual rights and societal values.

Cost Analysis and Feasibility of the Bullet Inversion System (BIS)

The Bullet Inversion System (BIS) represents a groundbreaking advance in personal protection technology. Its development and deployment involve intricate considerations related to cost, feasibility, and market potential. This analysis aims to explore the financial aspects of BIS, including the estimation of development and deployment costs, and the evaluation of its economic feasibility and market potential.

Estimating the Cost of BIS Development and Deployment

1. Research and Development (R&D) Costs

1.1. Conceptualization and Prototyping

The initial phase of BIS development involves significant R&D efforts. Costs in this phase include:

- **Feasibility Studies:** Assessing the theoretical and practical viability of BIS requires comprehensive studies involving expert consultations and preliminary simulations. These studies help in defining technical specifications and identifying potential challenges.
- **Prototype Development:** Building small-scale prototypes is a critical step. This includes the costs associated with materials, manufacturing, and testing. Advanced materials, such as superconductors and specialized alloys, can be expensive, and the prototyping process often involves iterative design improvements.

1.2. Advanced Technologies

BIS integrates advanced technologies such as high-energy magnetic fields and plasma generation. Costs include:

- **Material Costs:** High-performance materials like superconductors, advanced core materials for electromagnets, and specialized cooling systems contribute significantly to the development cost.
- **Technology Integration:** Incorporating cutting-edge technologies, such as plasma shields and electromagnetic induction systems, requires sophisticated engineering and integration efforts, which can be costly.

2. Manufacturing Costs

2.1. Production Scaling

Scaling up from prototypes to full-scale manufacturing involves various costs:

- **Manufacturing Setup:** Establishing production facilities or modifying existing ones to accommodate BIS manufacturing can be substantial. This includes the cost of machinery, production lines, and quality control systems.
- **Materials and Components:** Bulk purchasing of materials and components might reduce costs, but high-performance materials required for BIS still command a premium price. Additionally, ensuring consistency and quality in production adds to the cost.

2.2. Labor and Expertise

Skilled labor and specialized expertise are crucial:

- **Workforce Costs:** Hiring engineers, scientists, and technicians with expertise in electromagnetism, plasma physics, and materials science is essential. Labor costs, including salaries and training, are significant components of the overall cost.
- **Consulting and Licensing:** Engaging with experts for consultation and obtaining necessary patents and licenses also contribute to the cost.

3. Deployment Costs

3.1. Field Testing and Validation

Before deployment, BIS must undergo rigorous field testing:

- **Testing Infrastructure:** Establishing testing facilities and conducting field trials involve costs related to equipment, logistics, and safety measures. Ensuring that BIS performs effectively in real-world scenarios is crucial for its acceptance and reliability.
- **Compliance and Certification:** Meeting regulatory standards and obtaining certifications for safety and efficacy are mandatory steps. The cost of compliance, including testing and documentation, must be factored into the deployment budget.

3.2. Marketing and Distribution

Effective marketing and distribution strategies are essential for BIS adoption:

- **Marketing Campaigns:** Developing marketing strategies, creating promotional materials, and conducting outreach efforts to potential customers involve costs. Educating the market about BIS's benefits and features is crucial for its success.
- **Distribution Logistics:** Setting up distribution channels and managing logistics, including transportation and inventory management, adds to the deployment cost.

Evaluating Economic Feasibility and Market Potential

1. Cost-Benefit Analysis

1.1. Economic Value Proposition

Assessing BIS's economic feasibility involves evaluating its value proposition:

- **Cost Savings:** BIS technology may lead to long-term cost savings by reducing injuries, fatalities, and associated medical costs. Quantifying these savings provides a clearer picture of the economic benefits.
- **Risk Mitigation:** BIS could reduce risks for individuals and organizations, potentially lowering insurance premiums and liability costs. Analyzing these factors helps in understanding the broader economic impact.

1.2. Return on Investment (ROI)

Calculating the ROI for BIS involves comparing development and deployment costs with anticipated benefits:

- **Revenue Potential:** Estimating potential revenue from BIS sales, considering market size and pricing strategies, provides insights into financial returns. Identifying target markets, such as defense, law enforcement, and private security, helps in projecting revenue streams.
- **Payback Period:** Assessing the time required to recoup the initial investment through sales and other revenue sources is crucial for evaluating financial feasibility.

2. Market Potential and Demand

2.1. Market Research

Understanding market potential involves comprehensive market research:

- **Market Segmentation:** Identifying potential market segments, including defense contractors, security agencies, and high-risk individuals, helps in tailoring the BIS product to specific needs.
- **Competitive Analysis:** Analyzing competitors and existing technologies provides insights into BIS's competitive advantage. Understanding market trends, pricing strategies, and customer preferences is essential for positioning BIS effectively.

2.2. Adoption Barriers

Evaluating potential barriers to BIS adoption includes:

- **Regulatory Hurdles:** Identifying and addressing regulatory challenges related to BIS's deployment and use is crucial. Ensuring compliance with laws and standards helps in facilitating market entry.

- **Public Perception:** Understanding public perception and addressing concerns related to BIS, such as its safety and effectiveness, is vital for gaining acceptance. Educational efforts and transparent communication can mitigate skepticism.

3. Economic Impact and Sustainability

3.1. Long-Term Economic Impact

Assessing the long-term economic impact of BIS involves:

- **Sustainability:** Evaluating the sustainability of BIS technology in terms of environmental impact and resource utilization is important. Developing eco-friendly materials and manufacturing processes contributes to long-term feasibility.
- **Economic Growth:** Considering the potential for BIS to stimulate economic growth through job creation, technology innovation, and market expansion provides a broader perspective on its economic benefits.

3.2. Financial Viability

Ensuring the financial viability of BIS includes:

- **Funding and Investment:** Securing funding and investment for BIS development and deployment is crucial. Exploring sources of funding, such as venture capital, grants, and partnerships, helps in addressing financial needs.
- **Cost Management:** Implementing effective cost management strategies, including budgeting, cost control, and financial forecasting, ensures that BIS development and deployment stay within budget.

Conclusion

The development and deployment of the Bullet Inversion System (BIS) involve substantial costs and require careful economic analysis to ensure feasibility. Estimating costs includes research and development, manufacturing, and deployment expenses. Evaluating economic feasibility involves conducting a cost-benefit analysis, assessing market potential, and understanding adoption barriers. By addressing these factors, stakeholders can make informed decisions about the viability of BIS and its potential impact on personal protection technology.

Impact on Law Enforcement and Military: Potential Applications and Tactical Operations

The development of the Bullet Inversion System (BIS) introduces a transformative potential for law enforcement and military operations. By leveraging advanced electromagnetic technologies and plasma generation, BIS could significantly alter how tactical engagements are conducted, how personnel are protected, and how overall operational strategies are formulated. This exploration delves into the potential applications of BIS within these domains and assesses its impact on tactical operations.

Potential Applications in Law Enforcement and Military

1. Law Enforcement Applications

1.1. Enhanced Personal Protection

One of the primary applications of BIS in law enforcement is the enhancement of personal protection. The BIS could be integrated into body armor or tactical gear, providing officers with an additional layer of defense against ballistic threats. The ability to deflect or neutralize bullets in real-time offers substantial protection in high-risk scenarios, such as armed confrontations and hostage situations.

- **Active Threat Response:** In active shooter situations, BIS-equipped gear could protect officers, allowing them to focus on neutralizing threats rather than seeking cover. This added protection could potentially reduce the number of casualties and injuries among law enforcement personnel.
- **High-Risk Operations:** For SWAT teams and tactical units, BIS could be employed during high-stakes operations where exposure to gunfire is a significant risk. The system's ability to mitigate bullet impact enhances the safety and effectiveness of these specialized teams.

1.2. Vehicle Protection

BIS technology could also be applied to enhance vehicle protection. Law enforcement vehicles, such as armored cars or tactical response vehicles, could be equipped with BIS to protect against ballistic attacks. This application would be particularly relevant in environments where vehicle-based protection is critical, such as during high-risk transport or crowd control operations.

- **Armored Vehicle Enhancements:** Integrating BIS into the armor of law enforcement vehicles could improve their resilience against high-caliber ammunition, increasing the safety of personnel inside.
- **Public Safety Operations:** During public safety operations, such as crowd control or demonstrations, BIS-equipped vehicles could provide added protection for law enforcement officers in the event of armed confrontations.

2. Military Applications

2.1. Personal Armor and Gear

For military personnel, BIS offers potential advancements in personal armor and tactical gear. The system could be integrated into uniforms, helmets, and other protective gear, providing soldiers with enhanced defense against ballistic threats in combat zones.

- **Combat Situations:** In combat scenarios, BIS-equipped armor could protect soldiers from small arms fire and shrapnel, improving their survivability on the battlefield.

- **Special Operations:** Special operations forces, who often operate in high-risk environments, could benefit from BIS integration. The technology could enhance their protection during critical missions, such as direct action operations or counter-terrorism activities.

2.2. Equipment Protection

BIS technology could also be applied to protect military equipment and installations. The system's ability to deflect or neutralize bullets could be used to safeguard critical assets, such as command posts, communication equipment, and sensitive installations.

- **Base Defense:** BIS could be used in base defense systems to protect military facilities from ballistic threats. Integrating BIS into perimeter defenses or strategic installations could enhance security and reduce vulnerabilities.
- **Vehicle and Aircraft Protection:** Military vehicles and aircraft could be equipped with BIS to provide additional protection against attacks. This application would be especially relevant for high-value assets operating in hostile environments.

Assessing BIS Impact on Tactical Operations

1. Tactical Flexibility

The integration of BIS into law enforcement and military operations could significantly impact tactical flexibility:

1.1. Operational Adaptability

BIS technology could allow law enforcement and military units to adapt their tactics in response to evolving threats. For example, the ability to reduce the risk of injury from ballistic attacks could encourage more aggressive tactics, such as dynamic entry techniques or direct engagement with armed adversaries.

- **Increased Aggressiveness:** With enhanced protection, units might adopt more aggressive tactics, such as direct assaults on fortified positions or engagements with heavily armed opponents. This could shift the dynamics of tactical operations, potentially leading to faster resolution of high-risk situations.
- **Enhanced Confidence:** The presence of BIS could increase confidence among personnel, leading to more effective decision-making and operational execution. Knowing that they are protected from ballistic threats may encourage more decisive actions in critical scenarios.

1.2. Strategic Planning

BIS integration into tactical operations could influence strategic planning and mission execution:

- **Mission Planning:** The availability of BIS could lead to changes in mission planning and strategy. Planners might consider new tactics or operational approaches based on the enhanced protection offered by BIS, such as deploying units in more exposed positions or engaging threats at closer ranges.
- **Resource Allocation:** The integration of BIS could impact resource allocation decisions. Agencies and military units may prioritize investments in BIS technology for high-risk operations or critical assets, influencing overall budget and resource distribution.

2. Safety and Risk Management

2.1. Reducing Casualties

The primary impact of BIS on tactical operations is the potential reduction in casualties and injuries. By providing enhanced protection against ballistic threats, BIS technology could significantly decrease the number of personnel affected by gunfire during operations.

- **Casualty Reduction:** The ability to deflect or neutralize bullets could lead to a reduction in both fatal and non-fatal injuries among law enforcement and military personnel. This could improve overall mission effectiveness and reduce the burden on medical and support services.
- **Operational Continuity:** Reduced casualties could enhance operational continuity, allowing units to maintain their effectiveness and complete missions despite facing armed threats. This could be particularly important in prolonged or high-intensity operations.

2.2. Safety Protocols

The integration of BIS technology would also necessitate the development of new safety protocols and procedures:

- **Training and Integration:** Personnel would require training on the use and limitations of BIS-equipped gear. Understanding the technology's capabilities and limitations is crucial for effective integration into tactical operations.
- **Maintenance and Reliability:** Ensuring the reliability and maintenance of BIS technology is essential for operational success. Regular inspections and maintenance procedures would be necessary to ensure that BIS systems function as intended during high-stress situations.

Conclusion

The Bullet Inversion System (BIS) has the potential to revolutionize personal protection in law enforcement and military operations. Its applications in enhancing personal armor, vehicle protection, and equipment defense could significantly impact tactical operations, offering improved protection and flexibility. Assessing the BIS's impact involves understanding its influence on tactical decision-making, operational planning, and safety protocols. By integrating

BIS technology into their operations, law enforcement and military units could benefit from enhanced protection and operational effectiveness, potentially transforming how they approach high-risk scenarios and ensuring greater safety for personnel in the field.

User Training and Operation for Bullet Inversion System (BIS): Developing Programs and Ensuring Effective Use

The successful implementation of a Bullet Inversion System (BIS) relies not only on its technological prowess but also on the effective training of its users. Developing comprehensive training programs and ensuring the safe and efficient operation of BIS by non-experts are critical steps in translating technological advancements into practical, real-world applications. This exploration addresses the essential components of user training and operation for BIS, focusing on developing training programs and ensuring effective use by individuals with varying levels of expertise.

Developing Training Programs for BIS Users

1. Understanding BIS Technology

1.1. System Overview

A fundamental component of training programs is providing users with a thorough understanding of BIS technology. Training should begin with an overview of the BIS's core principles, including how it generates magnetic fields, the role of plasma shielding, and the system's operational mechanics.

- **Conceptual Foundation:** Users should be introduced to the theoretical principles underlying BIS, such as electromagnetism, plasma physics, and materials science. This foundation helps users grasp how the technology functions and why certain operational procedures are necessary.
- **System Components:** Training should cover the various components of the BIS, including magnetic field generators, plasma systems, and power supplies. Understanding each component's role and functionality is essential for effective operation and troubleshooting.

1.2. Practical Operation

Effective training must also include practical instruction on operating the BIS. This involves hands-on practice with the device, allowing users to familiarize themselves with its controls, settings, and operational procedures.

- **Device Handling:** Training should cover how to handle the BIS safely, including proper installation, calibration, and operation. Users should practice setting up the device, adjusting settings, and activating the system in controlled environments.

- **Simulation Exercises:** Simulated scenarios can help users practice operating the BIS under various conditions. Simulations allow users to experience potential challenges and develop problem-solving skills without the risks associated with real-world applications.

2. Safety Protocols and Emergency Procedures

2.1. Safety Training

Safety is a paramount concern when operating high-energy systems like the BIS. Training programs should include comprehensive safety protocols to protect users and bystanders.

- **Risk Awareness:** Users should be trained to recognize potential hazards associated with BIS operation, such as high electromagnetic fields, plasma generation, and high power consumption. Understanding these risks is crucial for preventing accidents and ensuring safe use.
- **Protective Measures:** Training should include instructions on using personal protective equipment (PPE) and implementing safety measures to mitigate risks. This may involve wearing appropriate gear, maintaining safe distances, and following operational guidelines to avoid exposure to hazardous conditions.

2.2. Emergency Procedures

Training programs must also cover emergency procedures for dealing with malfunctions, system failures, or accidental exposures.

- **Emergency Response:** Users should be trained to respond effectively to emergencies, such as system malfunctions or accidental exposure to high-energy fields. This includes knowing how to shut down the system safely, evacuating the area, and seeking medical assistance if needed.
- **Troubleshooting:** Training should provide guidance on basic troubleshooting techniques for common issues that may arise during BIS operation. Users should learn how to identify and address problems, such as system errors or performance issues, to minimize downtime and maintain operational effectiveness.

3. Advanced Training and Certification

3.1. Specialized Training

For advanced users or those involved in critical applications, specialized training may be required. This training focuses on in-depth knowledge and advanced operational techniques, including system customization, advanced troubleshooting, and integration with other technologies.

- **Custom Configuration:** Advanced training should cover how to customize BIS settings for specific applications or environments. Users should learn how to adjust parameters, optimize performance, and configure the system for different operational scenarios.

- **Integration Skills:** Training may also include instruction on integrating BIS with other technologies or protective systems. This involves understanding how BIS interacts with body armor, automated threat detection systems, or other defense mechanisms.

3.2. Certification

Certification programs can validate users' proficiency in operating BIS and ensure consistent standards across different applications.

- **Certification Process:** Certification programs typically involve testing users' knowledge and skills through practical assessments and theoretical examinations. Successful certification indicates that users are competent in operating BIS safely and effectively.
- **Ongoing Education:** Regular updates and refresher courses should be provided to keep users informed about technological advancements, changes in operational procedures, and emerging safety practices.

Ensuring Effective and Safe Operation by Non-Experts

1. User Interface Design

1.1. Intuitive Controls

Ensuring that BIS can be operated effectively by non-experts requires an intuitive user interface. The design of the control panel, display screens, and operational buttons should prioritize ease of use and clarity.

- **User-Friendly Design:** Controls should be straightforward and clearly labeled, with visual indicators and prompts to guide users through the operation process. Reducing complexity and minimizing the potential for user error is crucial for ensuring effective use.
- **Feedback Mechanisms:** The BIS interface should provide real-time feedback on system status, performance, and any potential issues. Clear visual or auditory signals can alert users to changes in system conditions or operational requirements.

1.2. Training Aids

Supplementary training aids, such as user manuals, instructional videos, and online resources, can support non-experts in understanding and operating BIS.

- **Comprehensive Documentation:** Detailed user manuals and guides should provide step-by-step instructions for operating the BIS, including setup, calibration, and troubleshooting. These resources should be accessible and easy to understand.

- **Interactive Resources:** Online tutorials, videos, and interactive training modules can enhance users' understanding and reinforce key concepts. These resources can be particularly useful for visual learners and those seeking additional guidance.

2. Support Systems

2.1. Technical Support

Providing ongoing technical support is essential for ensuring effective operation by non-experts. Access to knowledgeable support personnel can help users address issues, seek guidance, and resolve technical problems.

- **Help Desks:** A dedicated help desk or support team should be available to assist users with questions or concerns. Support personnel should be trained to provide clear, concise guidance and troubleshooting assistance.
- **Remote Assistance:** Remote assistance tools, such as video calls or remote diagnostics, can enable support personnel to assist users in real-time, regardless of their location. This can be particularly valuable for addressing complex issues or providing on-site support.

2.2. Maintenance and Updates

Regular maintenance and updates are necessary to ensure that BIS systems continue to operate effectively and safely.

- **Scheduled Maintenance:** Users should be informed about routine maintenance tasks, such as system inspections, calibration, and software updates. Regular maintenance helps prevent issues and ensures optimal performance.
- **Software Updates:** BIS systems may require software updates to address bugs, improve functionality, or incorporate new features. Users should be provided with clear instructions for installing updates and maintaining system security.

Conclusion

Developing effective training programs and ensuring the safe and efficient operation of the Bullet Inversion System (BIS) are critical for maximizing its benefits and minimizing risks. Comprehensive training should encompass an understanding of BIS technology, practical operation, safety protocols, and advanced skills. Ensuring that non-experts can operate BIS effectively involves intuitive user interface design, supplementary training aids, and robust support systems. By addressing these aspects, organizations can enhance the effectiveness of BIS technology and ensure its successful integration into practical applications.

Public Acceptance and Perception of Bullet Inversion Systems (BIS): Gauging Opinion and Building Trust

The introduction of innovative technologies like the Bullet Inversion System (BIS) often prompts significant public interest and scrutiny. As a sophisticated personal protection device designed to redirect incoming projectiles using advanced electromagnetic and plasma technologies, the BIS faces the dual challenge of proving its technological efficacy while gaining public acceptance. This discourse explores how to gauge public opinion on BIS technology and outlines strategies for building trust and fostering acceptance.

Gauging Public Opinion on BIS Technology

1. Understanding Public Concerns

1.1. Safety and Security

Public perception of BIS technology is deeply influenced by concerns over safety and security. Understanding these concerns is crucial for addressing fears and misconceptions. Key issues often include:

- **Operational Safety:** People may worry about the safety of using a device that generates strong magnetic fields and plasma. Concerns might include potential health effects from electromagnetic fields, risks associated with plasma generation, and the overall reliability of the technology.
- **Device Reliability:** Public confidence in BIS technology depends on its perceived reliability. Questions about the system's effectiveness in real-world scenarios, including its ability to accurately redirect bullets and its durability under stress, are central to public opinion.

1.2. Ethical and Social Implications

The BIS also raises ethical and social considerations that influence public acceptance. These include:

- **Privacy Concerns:** The use of advanced surveillance and detection technologies in conjunction with BIS could raise privacy concerns. Ensuring that BIS technology is used responsibly and does not infringe on individual rights is essential for public trust.
- **Impact on Law Enforcement:** The integration of BIS technology into public spaces or personal protection might affect the role and effectiveness of law enforcement. Public opinion may be shaped by how BIS interacts with existing security measures and its potential implications for law enforcement practices.

2. Methods for Gauging Public Opinion

2.1. Surveys and Polls

Conducting surveys and polls is an effective way to gauge public opinion on BIS technology. These tools can provide quantitative data on public attitudes, concerns, and acceptance levels.

- **Designing Surveys:** Surveys should be carefully designed to capture a broad range of opinions and address specific concerns related to BIS. Questions might focus on perceived safety, effectiveness, ethical considerations, and willingness to adopt the technology.
- **Analyzing Results:** Analyzing survey data can reveal trends and patterns in public opinion. This information is valuable for understanding the general sentiment towards BIS and identifying areas where further education or reassurance may be needed.

2.2. Focus Groups

Focus groups provide qualitative insights into public perceptions and attitudes. These discussions involve a small group of individuals who provide in-depth feedback on BIS technology.

- **Conducting Focus Groups:** Focus groups should be facilitated by skilled moderators who can guide discussions and probe deeper into participants' thoughts and feelings. This approach allows for a nuanced understanding of public concerns and the opportunity to address specific issues in real time.
- **Interpreting Feedback:** The feedback gathered from focus groups can help identify common themes, concerns, and misconceptions. This information can guide the development of communication strategies and public engagement efforts.

Strategies for Gaining Public Trust and Acceptance

1. Transparent Communication

1.1. Providing Clear Information

Transparency is crucial for gaining public trust. Providing clear, accurate information about BIS technology helps address concerns and build confidence.

- **Educational Campaigns:** Launching educational campaigns that explain the principles of BIS, its safety features, and its operational capabilities can help demystify the technology. These campaigns should use accessible language and visuals to make complex concepts understandable to a broad audience.
- **Addressing Misconceptions:** Proactively addressing common misconceptions and fears about BIS technology is essential. This might involve providing evidence-based responses to concerns about safety, effectiveness, and ethical implications.

1.2. Engaging with the Public

Engaging with the public through forums, town halls, and online platforms can foster dialogue and build trust.

- **Public Forums:** Hosting public forums or community meetings provides an opportunity for individuals to ask questions, express concerns, and receive direct responses from experts. This interactive approach helps build rapport and credibility.
- **Online Engagement:** Utilizing social media and online platforms to share information, answer questions, and engage with the public can reach a wider audience. Regular updates and interactive content can keep the public informed and engaged.

2. Demonstrating Effectiveness and Safety

2.1. Showcasing Real-World Applications

Demonstrating the practical benefits and effectiveness of BIS technology can help build confidence and acceptance.

- **Field Demonstrations:** Organizing live demonstrations or simulations of BIS technology in action can provide tangible evidence of its capabilities. These demonstrations should highlight the technology's effectiveness in real-world scenarios and its potential benefits for personal safety.
- **Case Studies:** Sharing case studies or testimonials from early adopters or pilot programs can provide real-world examples of BIS technology in use. Positive experiences and successful implementations can enhance public perception and build credibility.

2.2. Independent Testing and Certification

Third-party testing and certification can provide an additional layer of validation for BIS technology.

- **Certification Bodies:** Obtaining certifications from reputable organizations or regulatory bodies can reassure the public of BIS technology's safety and reliability. Certification processes typically involve rigorous testing and adherence to industry standards.
- **Independent Reviews:** Encouraging independent reviews and assessments by experts or organizations can provide unbiased evaluations of BIS technology. Positive reviews from credible sources can enhance public trust and acceptance.

3. Ethical and Social Responsibility

3.1. Ethical Use Guidelines

Establishing clear guidelines for the ethical use of BIS technology is important for addressing concerns and demonstrating responsibility.

- **Ethical Framework:** Developing an ethical framework for the deployment and use of BIS technology can help address concerns about privacy, security, and social impact.

This framework should outline principles for responsible use and ensure that the technology is applied in ways that respect individual rights and public welfare.

- **Accountability Measures:** Implementing accountability measures, such as oversight committees or review boards, can ensure that BIS technology is used ethically and transparently. These measures can provide mechanisms for addressing issues or concerns that arise during the technology's deployment.

3.2. Collaboration with Stakeholders

Collaborating with stakeholders, including community organizations, law enforcement, and advocacy groups, can help build support and address concerns.

- **Stakeholder Engagement:** Engaging with diverse stakeholders can provide valuable insights and feedback on BIS technology. This collaboration can help identify potential issues, build consensus, and ensure that the technology aligns with community values and needs.
- **Partnerships:** Forming partnerships with organizations that have a vested interest in public safety and technology can enhance credibility and support. These partnerships can facilitate knowledge sharing, joint initiatives, and mutual support for the adoption of BIS technology.

Conclusion

Public acceptance and perception of Bullet Inversion System (BIS) technology are influenced by a range of factors, including safety concerns, ethical considerations, and the perceived effectiveness of the technology. By employing strategies to gauge public opinion, transparently communicate information, demonstrate effectiveness and safety, and address ethical and social responsibility, developers and proponents of BIS can build trust and foster acceptance. Engaging with the public through educational campaigns, demonstrations, and stakeholder collaborations is key to ensuring that BIS technology is perceived positively and adopted effectively. Through these efforts, the potential benefits of BIS technology can be realized, contributing to enhanced personal protection and safety.

Environmental Impact and Sustainability of Bullet Inversion Systems (BIS): Analyzing Footprints and Exploring Sustainable Solutions

The development and deployment of Bullet Inversion Systems (BIS) represent a significant leap in personal protection technology. However, as with any advanced technology, understanding its environmental impact and pursuing sustainability are crucial for long-term viability. This discussion delves into the environmental footprint of BIS technology and explores potential avenues for incorporating sustainable materials and energy sources.

Analyzing the Environmental Footprint of BIS

1. Production and Manufacturing

1.1. Material Extraction and Processing

The production of BIS involves various materials, including metals for electromagnets, superconducting materials, and components for plasma generation. Each of these materials has an associated environmental footprint:

- **Metals and Alloys:** The extraction and processing of metals like iron, cobalt, and rare-earth elements for magnets require significant energy and can lead to habitat destruction, water contamination, and increased greenhouse gas emissions. The mining processes also often involve substantial land disturbance and ecosystem disruption.
- **Superconductors:** High-temperature superconductors (HTS) such as yttrium barium copper oxide (YBCO) and low-temperature superconductors (LTS) like niobium-titanium (NbTi) involve complex and energy-intensive production processes. The synthesis of these materials can contribute to environmental impacts, including chemical waste and energy consumption.

1.2. Energy Consumption During Manufacturing

The manufacturing of BIS components demands substantial energy, particularly for producing high-performance magnets and superconductors. This energy consumption can contribute to carbon emissions and environmental degradation if sourced from non-renewable energy sources.

- **Electromagnet Production:** The production of electromagnets requires significant electrical energy, particularly for winding coils and operating cooling systems. Efficient management of this energy is essential to minimize environmental impact.
- **Plasma Generation Systems:** Plasma generation also requires high-energy inputs, especially for ionizing gases and maintaining stable plasma fields. The energy intensity of these processes further contributes to the overall environmental footprint.

2. Operational Impact

2.1. Energy Consumption in Use

The operational phase of BIS involves the continuous generation of magnetic fields and, in some cases, plasma shields. The energy demands during this phase must be carefully managed to reduce the environmental impact.

- **Power Requirements:** The power requirements for generating strong magnetic fields and maintaining plasma shields are considerable. Effective energy management strategies and the use of renewable energy sources can help mitigate the environmental impact during operation.
- **Heat Generation:** High-energy systems generate substantial heat, which must be managed through cooling systems. The energy used for cooling can contribute to overall energy consumption and environmental impact.

2.2. Waste Generation

BIS systems may generate waste during their lifecycle, including electronic waste from outdated components, used materials, and potential byproducts from manufacturing processes.

- **Electronic Waste:** Components of BIS, such as control systems and sensors, may become obsolete or fail over time, leading to electronic waste. Proper disposal and recycling practices are necessary to minimize environmental harm.
- **Material Waste:** The production and assembly of BIS components may result in material waste. Implementing efficient manufacturing processes and recycling practices can help reduce waste generation.

Exploring Sustainable Materials and Energy Sources

1. Sustainable Materials

1.1. Eco-Friendly Alternatives

The development and adoption of sustainable materials can significantly reduce the environmental impact of BIS technology.

- **Recycled Metals and Alloys:** Utilizing recycled metals and alloys for manufacturing electromagnets and other components can reduce the need for virgin materials and lower environmental impact. Recycling reduces energy consumption and minimizes waste associated with material extraction.
- **Biodegradable Materials:** Exploring biodegradable or less environmentally harmful materials for components can reduce the long-term impact of BIS technology. Materials with lower environmental footprints during production and disposal contribute to overall sustainability.

1.2. Advanced Superconducting Materials

Research into new superconducting materials with improved properties and lower environmental impact is essential for enhancing sustainability.

- **High-Temperature Superconductors:** Developing new high-temperature superconductors that operate efficiently at higher temperatures can reduce the energy required for cooling. Innovations in superconducting materials can also lead to reduced environmental impact during production.
- **Low-Impact Manufacturing:** Advancements in manufacturing techniques for superconductors that minimize waste and energy consumption can further enhance sustainability. These improvements contribute to a lower environmental footprint.

2. Renewable Energy Sources

2.1. Powering BIS with Renewable Energy

Integrating renewable energy sources into the operation of BIS systems can help reduce the overall environmental impact.

- **Solar Energy:** Utilizing solar power to generate electricity for BIS components can reduce reliance on non-renewable energy sources. Solar panels can be integrated into the design of BIS systems to provide a sustainable energy supply.
- **Wind Energy:** Wind turbines can generate electricity to power BIS systems, particularly in areas with strong and consistent wind patterns. Wind energy offers a renewable and environmentally friendly alternative to traditional power sources.

2.2. Improving Energy Efficiency

Enhancing the energy efficiency of BIS systems can reduce the overall demand for power and minimize environmental impact.

- **Energy-Efficient Components:** Implementing energy-efficient components and technologies in BIS systems can lower energy consumption. This includes optimizing power supplies, cooling systems, and control mechanisms.
- **Smart Energy Management:** Incorporating smart energy management systems can optimize the use of available power, reduce waste, and improve overall efficiency. Advanced control systems can dynamically adjust energy usage based on operational needs.

3. Lifecycle Management

3.1. Sustainable Design Practices

Designing BIS systems with sustainability in mind involves considering the entire lifecycle of the technology, from production to disposal.

- **Modular Design:** Developing BIS systems with modular components can facilitate repair and upgrades, reducing the need for complete replacements and minimizing waste.
- **End-of-Life Management:** Implementing end-of-life management practices, such as recycling programs and safe disposal methods, can reduce the environmental impact of BIS systems. Ensuring that materials and components are properly handled at the end of their lifecycle contributes to overall sustainability.

3.2. Research and Development

Ongoing research and development efforts are essential for identifying and implementing sustainable practices in BIS technology.

- **Material Research:** Continued research into new materials with lower environmental impact and improved performance can drive sustainability in BIS systems. Innovations in material science contribute to more eco-friendly technologies.

- **Energy Solutions:** Advancements in energy storage and management solutions can enhance the sustainability of BIS systems. Exploring new energy technologies and integrating them into BIS design can reduce environmental impact.

Conclusion

The environmental impact and sustainability of Bullet Inversion Systems (BIS) are critical considerations for the future of this advanced technology. By analyzing the environmental footprint of BIS production, operation, and disposal, and exploring sustainable materials and energy sources, we can work towards reducing the overall impact. Implementing eco-friendly materials, utilizing renewable energy sources, and adopting sustainable design practices are essential for ensuring that BIS technology aligns with environmental stewardship goals. Through ongoing research and development, the integration of sustainable practices into BIS systems will contribute to a more responsible and environmentally conscious approach to personal protection technology.

Comparative Analysis with Existing Technologies: Evaluating Bullet Inversion Systems (BIS) Against Current Body Armor and Protective Gear

The advent of Bullet Inversion Systems (BIS) represents a paradigm shift in personal protection technology. To fully appreciate the potential impact of BIS, it is essential to conduct a comparative analysis with existing body armor and protective gear. This analysis involves evaluating BIS in relation to current technologies, highlighting its advantages, and addressing its limitations.

Comparing BIS with Current Body Armor and Protective Gear

1. Current Body Armor Technologies

1.1. Kevlar and Aramid Fibers

Body armor technology has traditionally relied on materials such as Kevlar and aramid fibers. Kevlar, a registered trademark of DuPont, is a high-strength synthetic fiber renowned for its ballistic resistance. The principal features of Kevlar body armor include:

- **Strength and Flexibility:** Kevlar fibers are woven into a fabric that is both strong and flexible. This combination provides effective protection against ballistic threats while allowing for comfortable wear.
- **Layered Design:** Modern Kevlar vests typically feature multiple layers of fabric that disperse and absorb the energy of incoming projectiles. This layered design enhances the vest's ability to stop bullets and reduce blunt force trauma.

1.2. Ceramic and Composite Plates

To provide additional protection, many body armor systems incorporate ceramic or composite plates. These plates are designed to defeat high-velocity projectiles and include:

- **Ceramic Plates:** Made from materials such as boron carbide or silicon carbide, ceramic plates are capable of stopping armor-piercing rounds. The hard surface of ceramic plates disperses the impact energy and minimizes penetration.
- **Composite Plates:** These plates combine ceramic materials with aramid fibers or other composite materials to offer a balance between weight and protection. They provide improved ballistic performance while maintaining a reasonable level of flexibility.

2. Advantages and Limitations of BIS

2.1. Advantages of BIS

2.1.1. Advanced Protective Capabilities

One of the primary advantages of BIS is its ability to offer protection beyond conventional armor. BIS employs magnetic fields and potentially plasma shields to counteract projectiles, providing several potential benefits:

- **Enhanced Defense:** BIS has the potential to redirect or neutralize incoming bullets using magnetic fields. This capability could offer superior protection against a range of ballistic threats, potentially reducing the risk of injury from both conventional and armor-piercing rounds.
- **Versatility:** By using electromagnetic fields and plasma shields, BIS could potentially be adapted to defend against different types of projectiles, including those that traditional body armor struggles to address.

2.1.2. Potential for Future Integration

BIS technology holds promise for integration with other advanced protective measures:

- **Complementary Systems:** BIS could be used in conjunction with traditional body armor to create a multi-layered defense system. This hybrid approach might offer enhanced protection by combining the strengths of both technologies.
- **Modular Design:** The potential for modular integration with automated threat detection systems could enhance the responsiveness and effectiveness of BIS. This integration could enable real-time adjustments to defensive measures based on detected threats.

2.2. Limitations of BIS

2.2.1. Technical and Practical Challenges

While BIS offers exciting possibilities, several technical and practical challenges must be addressed:

- **Power Requirements:** Generating and maintaining strong magnetic fields or plasma shields requires substantial power. Ensuring a reliable and compact power source for BIS remains a significant challenge, particularly for portable applications.
- **Material Constraints:** Developing materials capable of withstanding the high-energy environments involved in BIS operation is complex. Addressing material durability and performance issues is critical for ensuring the effectiveness and safety of BIS technology.

2.2.2. Complexity and Cost

The complexity and cost of BIS technology present additional limitations:

- **Design Complexity:** The integration of magnetic fields and plasma shields into a cohesive system involves complex design and engineering challenges. Achieving reliable and effective operation requires advanced research and development efforts.
- **Cost Considerations:** The development and deployment of BIS are likely to involve high costs due to the advanced materials, technologies, and power systems required. This cost could impact the accessibility and widespread adoption of BIS technology.

3. Comparative Analysis

3.1. Effectiveness

In terms of effectiveness, BIS offers a potentially superior level of protection by actively altering the trajectory of incoming projectiles. This active defense capability contrasts with the passive protection provided by traditional body armor, which primarily relies on absorbing and dispersing impact energy.

- **Active Defense:** BIS's ability to use magnetic fields or plasma shields represents a proactive approach to personal protection. This active defense could potentially offer greater protection against a wider range of threats compared to passive armor systems.
- **Impact Reduction:** Traditional body armor systems are designed to reduce blunt force trauma and prevent penetration. BIS, with its potential to redirect or neutralize projectiles, might reduce the overall impact force experienced by the wearer.

3.2. Comfort and Mobility

Traditional body armor systems, such as those made from Kevlar and composite materials, are designed with comfort and mobility in mind:

- **Flexibility:** Kevlar vests and composite plates are designed to offer a balance between protection and mobility. They allow for a range of motion and are often tailored for comfortable wear during extended periods.
- **Ergonomics:** The design of traditional body armor considers ergonomics to ensure that wearers can perform their duties without excessive discomfort or restriction.

3.3. Cost and Accessibility

Cost and accessibility are important factors in the comparative analysis of BIS and existing technologies:

- **Affordability:** Traditional body armor systems are well-established and widely available, with established production processes that help control costs. In contrast, BIS technology may involve higher costs due to its advanced materials and power requirements.
- **Production Scale:** The scalability of BIS production is another consideration. Traditional body armor benefits from established manufacturing and distribution networks, while BIS technology may require new production and supply chain infrastructure.

4. Future Directions

4.1. Technological Integration

Future developments in BIS technology could involve integration with other advanced systems to address current limitations:

- **Hybrid Systems:** Combining BIS with existing armor technologies could create hybrid systems that leverage the strengths of both approaches. Research into such integrations could lead to more effective and versatile protective solutions.
- **Advancements in Power and Materials:** Continued research into power sources and materials for BIS could address current technical and practical challenges, improving the feasibility and performance of BIS technology.

4.2. Market Adoption

Increasing the accessibility and affordability of BIS technology could drive wider adoption:

- **Cost Reduction:** Innovations in manufacturing and materials could reduce the cost of BIS technology, making it more accessible to a broader range of users.
- **Market Integration:** Collaboration with existing body armor manufacturers and distributors could facilitate the integration of BIS into the market, ensuring that the technology reaches its potential users.

Conclusion

The comparative analysis of Bullet Inversion Systems (BIS) with current body armor and protective gear highlights both the potential advantages and limitations of this advanced technology. BIS offers promising advancements in active defense capabilities and integration potential but faces challenges related to power requirements, complexity, and cost. By addressing these challenges and leveraging the strengths of existing technologies, BIS has the potential to enhance personal protection in innovative and effective ways. Continued research, development, and market integration will be key to realizing the full potential of BIS technology and advancing the future of personal protection.

Potential for Non-Lethal Applications of Bullet Inversion Systems (BIS)

The development of Bullet Inversion Systems (BIS) promises to revolutionize not only personal protection against lethal threats but also the realm of non-lethal applications. By harnessing advanced technologies such as electromagnetic fields and plasma shields, BIS has the potential to be adapted for various non-lethal uses. This exploration focuses on the possible applications of BIS technology in crowd control and personal safety, illustrating its versatility and broader societal impact.

Exploring Non-Lethal Uses for BIS Technology

1. Non-Lethal Crowd Control

1.1. Enhanced Safety and Reduced Harm

Crowd control situations often require strategies that balance effective management with the minimization of harm to individuals. BIS technology, with its ability to deflect or neutralize projectiles without causing physical injury, could offer new methods for non-lethal crowd control:

- **Electromagnetic Barriers:** By deploying magnetic fields to disrupt or divert projectiles, BIS technology could be adapted to create barriers that prevent the escalation of violent confrontations. This technology might deter individuals from using projectile-based weapons, reducing the risk of injury in volatile situations.
- **Directed Energy Shields:** Plasma shields could be used to create non-lethal barriers or deterrents. For example, a plasma shield might be employed to create a protective zone that deflects thrown objects or projectiles, thereby reducing the potential for harm to bystanders and law enforcement personnel.

1.2. Minimizing Physical Confrontations

In scenarios where physical confrontations are likely, BIS technology could assist in de-escalating tensions:

- **Active Defense Systems:** BIS could be integrated into crowd control equipment to provide real-time protection for law enforcement officers and security personnel. By using magnetic fields to neutralize thrown objects or projectiles, BIS technology could help prevent direct physical engagement and reduce the need for more aggressive tactics.
- **Non-Lethal Projectiles:** The technology might be adapted to intercept and neutralize non-lethal projectiles, such as bean bags or rubber bullets. By redirecting or dissipating these projectiles, BIS could reduce the risk of injury associated with their use.

2. Personal Safety Applications

2.1. Personal Protection in High-Risk Situations

The non-lethal potential of BIS technology extends to personal safety applications, offering protection in various high-risk situations:

- **Personal Defense Shields:** BIS technology could be incorporated into wearable devices or personal shields. These devices would provide individuals with an additional layer of protection by using electromagnetic fields to deflect or neutralize incoming projectiles. Such a system would be valuable for personal security in areas with heightened risk of violence.
- **Home Security Systems:** BIS technology could be integrated into home security systems to protect against break-ins and other threats. For example, magnetic fields could be used to deter intruders or to neutralize projectiles fired during a home invasion.

2.2. Enhancing Safety in Public Events

Public events often involve large crowds and increased risk of conflict or violence. BIS technology could enhance safety measures in such settings:

- **Event Security Integration:** By incorporating BIS technology into security measures for public events, organizers could enhance the protection of attendees and staff. Electromagnetic fields or plasma shields could be used to create safe zones or to deflect potential projectiles, reducing the risk of injury.
- **Emergency Response:** In emergencies or active shooter situations, BIS technology could be deployed as part of emergency response strategies. For example, rapidly deployable BIS systems could provide temporary protection for individuals in critical situations, allowing for evacuation or intervention.

Theoretical, Practical, and Logistical Considerations

1. Theoretical Foundations

The non-lethal applications of BIS technology are grounded in the theoretical principles of electromagnetism and plasma physics. Understanding how electromagnetic fields interact with materials and how plasma shields can be stabilized is crucial for developing effective non-lethal systems:

- **Magnetic Field Interactions:** Theoretical models of magnetic field interactions with projectiles provide insight into how BIS technology can be used to create barriers or deflect objects. Research into the optimal strength and configuration of magnetic fields is essential for effective non-lethal applications.
- **Plasma Shield Stability:** Theoretical studies on plasma physics help in understanding how plasma shields can be created and maintained. This knowledge is crucial for designing stable and effective non-lethal barriers.

2. Practical Implementation

Practical implementation of BIS technology for non-lethal applications involves several key considerations:

- **Device Design:** Designing BIS systems for non-lethal applications requires careful consideration of form factors and usability. Devices must be lightweight, portable, and user-friendly to be effective in real-world scenarios.
- **Power and Energy:** Non-lethal BIS systems must be designed with energy efficiency in mind. Compact and reliable power sources are essential for ensuring the effectiveness and operational reliability of these systems.

3. Logistical Challenges

Logistical challenges associated with the deployment of BIS technology for non-lethal applications include:

- **Cost and Accessibility:** The development and deployment of BIS systems involve significant costs. Ensuring that these systems are cost-effective and accessible for various applications is critical for widespread adoption.
- **Training and Maintenance:** Effective use of BIS technology requires proper training and maintenance. Ensuring that users are well-trained in the operation of BIS systems and that the systems are maintained in good working condition is essential for successful deployment.

4. Future Directions

The future of BIS technology in non-lethal applications involves ongoing research and development:

- **Material and Technology Advancements:** Continued research into advanced materials and technologies will improve the performance and effectiveness of BIS systems. Innovations in electromagnetic materials and plasma generation will drive future developments.

- **Integration with Other Systems:** Exploring how BIS technology can be integrated with other security and safety systems will enhance its effectiveness. For example, combining BIS with automated threat detection systems could provide a comprehensive approach to non-lethal protection.

- **Public Acceptance and Regulation:** Ensuring that BIS technology is publicly accepted and regulated appropriately is crucial for its successful implementation. Engaging with stakeholders and addressing regulatory concerns will help facilitate the adoption of BIS technology in non-lethal applications.

Conclusion

The potential for non-lethal applications of Bullet Inversion Systems (BIS) is significant, offering new possibilities for crowd control, personal safety, and public event security. By leveraging advanced electromagnetic fields and plasma shields, BIS technology can provide innovative solutions to reduce harm and enhance safety. Theoretical insights, practical considerations, and logistical challenges must be addressed to realize the full potential of BIS technology in non-lethal applications. Continued research, development, and public engagement will be key to advancing these technologies and ensuring their effective and responsible use.

Case Studies and Historical Precedents in Protective Technologies

The exploration of Bullet Inversion Systems (BIS) for personal and public safety benefits from a robust understanding of historical attempts at similar technologies. Analyzing past research, development efforts, and technological innovations provides valuable insights into the feasibility and challenges of such advanced systems. This exploration focuses on historical precedents in protective technologies, examining notable case studies, and distilling lessons learned from past endeavors.

Historical Attempts at Similar Technologies

1. Early Bullet-Deflection Technologies

1.1. Ancient and Medieval Armor

Historically, various forms of armor were developed to protect individuals from projectiles and melee weapons. For example:

- **Medieval Plate Armor:** During the medieval period, plate armor was designed to deflect arrows and bolts. The design focused on creating a smooth, hard surface to disperse the impact force of projectiles.
- **Chainmail:** This early form of body armor, made of interlinked metal rings, provided protection against cutting attacks and some penetration from arrows, though it was less effective against high-velocity projectiles.

These early technologies laid the groundwork for understanding material properties and the principles of deflecting projectiles, but they were limited by the available materials and manufacturing techniques of the time.

1.2. 20th Century Developments

The 20th century saw significant advancements in protective technologies, including attempts at active protection systems:

- **Active Protection Systems (APS) for Vehicles:** Developed during the latter half of the 20th century, APS technologies were designed to protect military vehicles from incoming projectiles. These systems used radar and sensors to detect threats and deploy countermeasures, such as explosive or non-explosive interceptors, to neutralize incoming projectiles.

- **Bulletproof Vests:** Modern bulletproof vests, made from materials like Kevlar and ceramics, represent significant advancements in personal protective equipment. They work by distributing the force of a bullet over a larger area and absorbing the impact, though they do not deflect projectiles as BIS technology aims to do.

2. Lessons Learned from Past Research and Development

2.1. Material Limitations

A recurring theme in the development of protective technologies is the limitation of materials:

- **Material Durability:** Many early protective technologies faced challenges related to material durability. For example, medieval armor was effective in deflecting certain projectiles but could be penetrated by others, depending on the force and type of the weapon.
- **Weight and Mobility:** Modern body armor, while effective, often struggles with issues related to weight and mobility. Heavy materials can limit the wearer's movement, highlighting the need for innovations that balance protection with comfort and practicality.

2.2. Technological Constraints

Technological constraints have also influenced the development of protective systems:

- **Energy and Power Requirements:** Early active protection systems and other advanced technologies often faced challenges related to power requirements and energy efficiency. APS systems required significant energy to operate sensors and countermeasures, which influenced their practical deployment.
- **Complexity and Cost:** The complexity of integrating advanced technologies with existing systems has historically been a challenge. For example, APS systems for vehicles are complex and costly, limiting their widespread adoption outside military applications.

2.3. Effectiveness and Reliability

Ensuring the effectiveness and reliability of protective technologies is crucial:

- **Testing and Validation:** Historical attempts often revealed gaps in testing and validation. For instance, early bulletproof vests required extensive testing to ensure they provided adequate protection under various conditions.
- **Operational Performance:** The real-world performance of protective systems can differ significantly from laboratory conditions. Early systems sometimes failed to meet expectations in actual combat or public safety scenarios.

3. Contemporary Innovations and Insights

3.1. Advances in Material Science

Recent advancements in materials science have addressed some of the limitations observed in historical technologies:

- **Advanced Composites:** Modern materials such as advanced composites and nanomaterials offer improved protection while reducing weight. These materials are increasingly used in body armor and vehicle protection systems.
- **Superconductors and Electromagnets:** Innovations in superconducting materials and electromagnets offer new possibilities for high-performance protective technologies. For instance, superconductors can generate powerful magnetic fields with minimal energy loss, which is relevant for BIS technology.

3.2. Integration with Modern Technologies

The integration of advanced technologies has also influenced the development of protective systems:

- **Smart Systems and AI:** The integration of smart systems and artificial intelligence (AI) enhances the functionality of protective technologies. For example, modern APS systems use AI to analyze threats and deploy countermeasures more effectively.
- **Miniaturization and Portability:** Advances in miniaturization and portable power sources have made it feasible to develop compact and effective protective devices. This is particularly relevant for BIS technology, which requires compact and efficient systems for practical deployment.

4. Implications for Future Development

4.1. Addressing Historical Challenges

The development of BIS technology can benefit from addressing the challenges encountered in historical attempts:

- **Material Innovations:** Continued research into new materials with superior properties will address issues of durability and performance. Innovations in energy-efficient materials and lightweight composites will enhance the feasibility of BIS systems.
- **Technological Integration:** Effective integration of BIS technology with existing systems will require overcoming challenges related to power, complexity, and cost. Collaboration between material scientists, engineers, and technology developers will be crucial.

4.2. Enhancing Testing and Validation

Improving testing and validation processes will ensure the reliability and effectiveness of BIS systems:

- **Comprehensive Testing:** Rigorous testing under various conditions will be essential to validate the performance of BIS technology. This includes field testing, simulation, and scenario-based evaluations.
- **User Feedback:** Incorporating user feedback and real-world performance data will help refine BIS technology and address any limitations or issues identified during deployment.

Conclusion

The study of historical attempts at protective technologies provides valuable insights into the development of Bullet Inversion Systems (BIS). By examining early bullet-deflection technologies, active protection systems, and contemporary innovations, we gain a deeper understanding of the challenges and opportunities in developing advanced protective systems. Lessons learned from past research and development efforts underscore the importance of material durability, technological integration, and effective testing. As we move forward, leveraging these insights will be crucial in advancing BIS technology and achieving its potential in personal and public safety.

Technological Challenges and Overcoming Them in Bullet Inversion Systems (BIS)

Developing a Bullet Inversion System (BIS) presents numerous technological challenges that need to be addressed for the successful realization of such a sophisticated device. Identifying these key technical obstacles and proposing viable solutions is essential for advancing this technology. This discussion explores the significant challenges associated with BIS development and outlines potential pathways to overcome these obstacles.

Identifying Key Technical Obstacles

1. Magnetic Field Generation and Control

1.1. Strength and Stability

A primary challenge in BIS technology is generating a magnetic field strong enough to deflect or alter the trajectory of high-velocity bullets. The strength of the magnetic field needed to affect a bullet—typically composed of non-ferromagnetic materials like lead—is substantial. Maintaining field stability while achieving sufficient strength adds another layer of complexity.

1.2. Precision and Control

Controlling the magnetic field with precision is crucial. Variations in field strength and orientation can affect the system's ability to consistently and reliably alter a bullet's path. The challenge lies in dynamically adjusting the magnetic field to respond to real-time threats.

2. Energy Requirements and Efficiency

2.1. Power Consumption

Generating strong magnetic fields requires significant energy. The power consumption associated with maintaining high-intensity fields is a critical challenge. High-energy systems can be cumbersome and impractical if not managed efficiently.

2.2. Battery Life and Storage

The power supply for BIS needs to be both robust and long-lasting. Current battery technologies may not offer the necessary energy density and longevity for sustained operation. Additionally, the size and weight of batteries must be balanced against the device's portability and practical use.

3. Material Durability and Performance

3.1. Heat Generation

High-intensity magnetic fields and the associated power requirements lead to substantial heat generation. The materials used in BIS must withstand these thermal stresses without degrading. Effective heat management solutions are needed to ensure that materials retain their performance under operational conditions.

3.2. Wear and Tear

Over time, exposure to high-energy environments can lead to material degradation. Components must be resilient to continuous operational demands and environmental factors, which can affect their longevity and reliability.

4. Miniaturization and Portability

4.1. Compact Design

Designing a BIS that is both powerful and compact poses a significant challenge. Miniaturizing magnetic field generators and power systems without compromising performance requires advanced engineering and materials science.

4.2. Integration

Integrating all components of the BIS into a single, portable unit involves complex engineering challenges. Ensuring that all elements function cohesively while maintaining user-friendly dimensions and weight is crucial for practical deployment.

Proposed Solutions and Pathways to Overcome Challenges

1. Advances in Magnetic Field Generation

1.1. Superconducting Materials

Utilizing superconducting materials can significantly enhance magnetic field strength and stability. Superconductors, such as high-temperature superconductors (HTS) like yttrium barium copper oxide (YBCO), can generate powerful magnetic fields with minimal energy loss. Innovations in cooling technologies to maintain superconductors at operational temperatures are critical to this approach.

1.2. Advanced Electromagnetic Designs

Developing new designs for electromagnetic coils and magnetic field generators, such as using superconducting solenoids or toroidal configurations, can improve field strength and control. Precision engineering and advanced manufacturing techniques, like 3D printing, can create intricate designs for enhanced performance.

2. Enhancing Energy Efficiency

2.1. Improved Battery Technologies

Research into new battery technologies, such as lithium-sulfur or solid-state batteries, could provide higher energy densities and longer life spans. These technologies offer the potential for smaller, lighter power sources with improved performance characteristics suitable for BIS applications.

2.2. Supercapacitors

Supercapacitors, with their ability to store and release large amounts of energy rapidly, can complement batteries by providing bursts of high power. Integrating supercapacitors with advanced battery systems could enhance overall energy management and efficiency for BIS.

3. Addressing Material Durability

3.1. Advanced Cooling Systems

Implementing advanced cooling systems, such as phase-change materials or active liquid cooling, can manage heat generated by high-intensity magnetic fields. Efficient thermal management systems will prevent overheating and material degradation, ensuring sustained performance.

3.2. High-Performance Materials

Developing new materials with higher thermal resistance and durability will address the issue of wear and tear. Nanomaterials and advanced composites could provide improved resistance to thermal and mechanical stresses, extending the lifespan of BIS components.

4. Miniaturization and Design Optimization

4.1. Precision Engineering

Advanced manufacturing techniques, including microfabrication and additive manufacturing, enable the creation of compact and precise components. These techniques allow for the integration of complex designs into small, efficient packages.

4.2. Modular Design

Adopting a modular design approach can facilitate the integration of different components into a unified system. Modular designs allow for flexibility in component selection and easy upgrades or replacements, which can enhance the system's overall performance and adaptability.

5. Integration and Testing

5.1. Integrated Systems Testing

Simulating and testing integrated systems in various conditions will provide insights into their performance and reliability. Comprehensive testing protocols, including field trials and stress testing, will help refine designs and identify any potential issues before full-scale deployment.

5.2. Feedback Loops

Incorporating feedback from real-world use and iterative design improvements will ensure that the BIS system evolves to meet practical requirements effectively. Continuous feedback and adaptation will be essential for overcoming unforeseen challenges and optimizing performance.

Conclusion

Developing a Bullet Inversion System (BIS) involves navigating a range of technological challenges, from generating and controlling powerful magnetic fields to addressing energy requirements, material durability, and miniaturization. By leveraging advances in superconducting materials, energy storage technologies, and precision engineering, it is possible to overcome these obstacles and create a functional and practical BIS. Addressing these challenges through innovative solutions and rigorous testing will pave the way for the successful implementation of BIS technology, potentially revolutionizing personal and public safety systems.

Collaborative Research Opportunities in Bullet Inversion Systems (BIS)

Collaborative research plays a pivotal role in advancing complex technologies such as the Bullet Inversion System (BIS). The interdisciplinary nature of BIS development necessitates partnerships between universities, research institutions, and industry partners. This exploration highlights the potential for collaboration, the benefits of engaging with various stakeholders, and the strategies for effective research and development (R&D) in the context of BIS technology.

Potential for Collaboration with Universities and Research Institutions

1. Leveraging Academic Expertise

Universities and research institutions are hubs of advanced knowledge and innovation, offering expertise in various fields essential for BIS development. Collaboration with these entities can provide access to cutting-edge research in electromagnetism, materials science, plasma physics, and energy storage technologies. Academic institutions often possess specialized equipment and facilities for high-level research, including advanced simulation tools, laboratory setups, and high-precision measurement instruments.

1.1. Research Projects and Grants

Engaging with universities allows for the initiation of joint research projects and the application for funding through grants and government programs. These collaborative projects can focus on specific aspects of BIS development, such as improving magnetic field generation, enhancing material durability, or developing energy-efficient solutions. Universities often have experience in securing research grants and can assist in navigating funding opportunities.

1.2. Access to Research Facilities

University laboratories and research centers are equipped with specialized tools and technologies not always available in private industry. Collaborating with academic institutions provides access to these resources, including high-field magnets, advanced imaging systems, and sophisticated computational platforms. This access accelerates the development process by allowing for more extensive experimentation and validation.

2. Interdisciplinary Research Opportunities

BIS technology intersects with multiple scientific disciplines, including physics, engineering, and material science. Collaborative research with universities enables the formation of interdisciplinary teams that bring together experts from diverse fields. This collaboration fosters innovation by integrating different perspectives and approaches.

2.1. Joint Research Centers

Establishing joint research centers between industry and academia can focus on BIS-related topics. These centers facilitate continuous collaboration, knowledge exchange, and resource sharing. They also provide a platform for conducting large-scale experiments and field trials, which are crucial for testing and validating BIS technologies.

2.2. Academic Publications and Conferences

Collaborating with universities often leads to academic publications and presentations at conferences. These activities not only disseminate research findings but also enhance the visibility of BIS technology in the scientific community. Engaging with academics through publications and conferences can attract additional collaborators and funding sources.

Engaging with Industry Partners for R&D

1. Industry Collaboration Benefits

Partnering with industry leaders and technology firms provides practical insights and accelerates the development and commercialization of BIS technology. Industry partners bring real-world experience, market knowledge, and technical expertise that complement academic research.

1.1. Technological Integration

Industry partners can offer valuable input on integrating BIS technology into existing systems and products. Their experience in developing and deploying advanced technologies helps ensure that BIS systems are practical, scalable, and aligned with market needs. This collaboration can lead to the creation of prototypes and pilot projects that test BIS technology in real-world scenarios.

1.2. Market Insights and Commercialization

Industry partners have a deep understanding of market trends and consumer needs. Collaborating with them provides insights into the commercial viability of BIS technology and helps in developing strategies for market entry and expansion. Industry partners can assist in navigating regulatory requirements, identifying potential customers, and creating effective marketing strategies.

2. Collaborative R&D Models

2.1. Public-Private Partnerships (PPPs)

Public-private partnerships (PPPs) are effective models for collaborative R&D. In these arrangements, public institutions (such as universities and government agencies) and private companies work together to advance technology development. PPPs can leverage public funding and private sector expertise to address specific research challenges and accelerate innovation.

2.2. Industry-Academia Consortia

Forming consortia involving multiple industry and academic partners can pool resources, share risks, and enhance research capabilities. Consortia can focus on specific aspects of BIS technology, such as energy efficiency or material science. This collaborative approach fosters knowledge sharing and facilitates the development of comprehensive solutions.

2.3. Technology Incubators and Accelerators

Technology incubators and accelerators provide support for early-stage innovations, including BIS technology. These programs offer resources such as funding, mentorship, and access to networks of experts and potential investors. Engaging with incubators and accelerators can help refine BIS prototypes, develop business models, and accelerate the path to commercialization.

3. Strategies for Effective Collaboration

3.1. Establish Clear Objectives

Setting clear research objectives and defining roles and responsibilities are crucial for successful collaboration. Establishing shared goals and aligning expectations ensure that all partners work towards common objectives and contribute effectively to the research and development process.

3.2. Foster Open Communication

Effective communication is key to successful collaboration. Regular meetings, progress reports, and open channels for discussion help address challenges, share insights, and make informed decisions. Building strong relationships between partners enhances coordination and collaboration.

3.3. Protect Intellectual Property

Collaborative research often involves sharing intellectual property (IP) and proprietary information. Establishing clear agreements on IP rights, data sharing, and confidentiality is essential to protect the interests of all parties involved. Negotiating these agreements upfront helps avoid conflicts and ensures fair distribution of benefits.

3.4. Evaluate and Adapt

Continuous evaluation of the collaboration process and outcomes allows for adjustments and improvements. Regular assessments of research progress, resource allocation, and partnership dynamics help optimize the effectiveness of the collaboration. Adapting strategies based on feedback and results ensures that the partnership remains productive and focused on achieving research goals.

Conclusion

Collaborative research is essential for advancing complex technologies such as Bullet Inversion Systems (BIS). Engaging with universities and research institutions provides access to cutting-edge knowledge, specialized resources, and interdisciplinary expertise. Industry partnerships offer practical insights, market knowledge, and commercialization support. Effective collaboration involves clear objectives, open communication, and protection of intellectual property. By leveraging the strengths of academic and industry partners, collaborative research can accelerate the development of BIS technology and drive innovation in personal and public safety systems.

Government and Military Funding for Bullet Inversion Systems (BIS)

Securing funding from government and military sources is crucial for the development of advanced technologies such as the Bullet Inversion System (BIS). This funding supports research, prototyping, testing, and eventual deployment, playing a key role in transforming theoretical concepts into practical, real-world applications. This discussion explores the various funding opportunities available, strategies for grant applications, and proposal techniques essential for obtaining financial support from government and military sources.

Exploring Funding Opportunities from Government and Military Sources

1. Government Funding Programs

Government agencies at both national and local levels offer various funding programs to support technological innovation and research. These programs are designed to advance national interests, promote scientific research, and foster technological development.

1.1. Research Grants and Fellowships

Government grants, such as those from the National Science Foundation (NSF), the Department of Energy (DOE), and the National Institutes of Health (NIH), provide financial support for research and development. These grants often target specific areas of technology and innovation, including advanced materials, electromagnetism, and energy systems relevant to BIS. Researchers can apply for grants to fund exploratory research, feasibility studies, and prototype development.

1.2. Defense Research and Development Programs

The Department of Defense (DoD) and related military agencies offer funding through programs like the Defense Advanced Research Projects Agency (DARPA) and the Office of Naval Research (ONR). These programs are aimed at advancing cutting-edge technologies with potential military applications. BIS technology aligns with the defense sector's interests in personal protection, making it a suitable candidate for funding under such programs.

1.3. Innovation and Technology Transfer Initiatives

Government initiatives, such as the Small Business Innovation Research (SBIR) and Small Business Technology Transfer (STTR) programs, support the commercialization of innovative technologies. These programs offer funding to small businesses and research institutions that collaborate to develop and transition new technologies into practical applications. BIS development could benefit from these initiatives, especially in the early stages of technology maturation.

2. Military Funding Opportunities

Military funding supports technologies that enhance national security, improve soldier safety, and advance defense capabilities. The military often funds research that aligns with its strategic objectives and operational needs.

2.1. Research Contracts and Cooperative Agreements

The military provides funding through research contracts and cooperative agreements, which involve partnerships between military agencies and research entities. These agreements can support the development of BIS technology, including aspects such as electromagnetic field generation and material science. Contracts often specify performance milestones and deliverables, ensuring that funded research meets military requirements.

2.2. Technology Demonstration and Evaluation Programs

The military conducts technology demonstration and evaluation programs to assess the practicality and effectiveness of emerging technologies. Funding for BIS technology could be obtained through these programs by demonstrating the system's potential benefits for military applications, such as enhanced personal protection and improved safety in combat situations.

2.3. Innovation and Acceleration Programs

Military innovation programs, such as the Army's Modernization Strategy and the Air Force's AFWERX initiative, focus on accelerating the development and deployment of new technologies. These programs provide funding and resources to rapidly prototype and field-test advanced systems. BIS technology could gain support from these programs by aligning with military modernization goals and demonstrating its relevance to current defense needs.

Grant Applications and Proposal Strategies

1. Understanding the Funding Landscape

To effectively secure funding, it is essential to understand the funding landscape, including the goals, priorities, and application processes of various government and military programs. Thorough research into the specific funding opportunities available, their eligibility criteria, and their alignment with BIS technology is crucial for crafting successful proposals.

1.1. Identifying Relevant Funding Sources

Identifying the most relevant funding sources involves researching agencies and programs that focus on technologies related to BIS. This includes examining the specific objectives of government and military programs, such as advancements in personal protection, electromagnetic technologies, and energy systems. Understanding these objectives helps tailor proposals to meet the specific interests of the funding entities.

1.2. Analyzing Proposal Requirements

Each funding program has unique requirements and guidelines for proposal submission. Analyzing these requirements, including application formats, documentation, and evaluation criteria, ensures that proposals are compliant and well-prepared. Attention to detail in adhering to these guidelines is critical for increasing the likelihood of securing funding.

2. Crafting a Compelling Proposal

A compelling proposal clearly articulates the significance, innovation, and impact of BIS technology. It should address the specific interests of the funding agency, demonstrate the potential benefits of the technology, and provide a detailed plan for development and implementation.

2.1. Defining the Problem and Objectives

The proposal should begin by defining the problem BIS technology addresses and outlining the project's objectives. This includes explaining the need for BIS technology, its potential impact on personal protection, and how it aligns with the funding agency's priorities. Clearly stating the problem and objectives sets the foundation for the proposal.

2.2. Presenting a Robust Research Plan

A robust research plan details the methodology, milestones, and expected outcomes of the project. This plan should include technical approaches, project timelines, and resource requirements. Demonstrating a well-thought-out plan with achievable milestones and clear deliverables reassures funders of the project's feasibility and potential for success.

2.3. Highlighting Innovation and Impact

Highlighting the innovative aspects of BIS technology and its potential impact is crucial for capturing the interest of funding agencies. The proposal should emphasize the uniqueness of the technology, its potential benefits for national security or public safety, and its alignment with the funding agency's mission. Providing evidence of innovation and impact strengthens the proposal's appeal.

2.4. Detailing Budget and Resource Allocation

The budget section of the proposal should provide a detailed breakdown of costs, including research, development, materials, and personnel. Justifying expenses and demonstrating efficient use of resources is essential for gaining the confidence of funders. Transparent and realistic budgeting ensures that the project's financial needs are clearly communicated.

2.5. Demonstrating Expertise and Capability

Showcasing the expertise and capabilities of the research team is crucial for establishing credibility. This includes presenting the qualifications of key personnel, their relevant experience, and their track record in similar projects. Demonstrating a strong research team reinforces the proposal's viability and potential for successful outcomes.

3. Submitting and Following Up

Submitting the proposal involves adhering to the deadlines and submission procedures specified by the funding agency. After submission, following up with the agency to confirm receipt and address any additional queries is essential for maintaining communication and ensuring a smooth review process.

3.1. Addressing Feedback and Revisions

If feedback is provided by the funding agency, addressing it promptly and making necessary revisions enhances the proposal's chances of approval. Constructive feedback provides valuable insights for improving the proposal and aligning it more closely with the agency's expectations.

3.2. Building Relationships with Funders

Building relationships with funding agencies through regular communication, updates on project progress, and participation in relevant events fosters positive engagement. Strong relationships can facilitate future funding opportunities and collaborations.

Conclusion

Government and military funding plays a critical role in advancing technologies such as the Bullet Inversion System (BIS). Understanding the various funding opportunities available, including research grants, defense contracts, and innovation programs, is essential for securing financial support. Crafting compelling proposals that clearly articulate the significance, innovation, and impact of BIS technology, while adhering to proposal guidelines and budgeting requirements, enhances the likelihood of obtaining funding. Effective collaboration with funding agencies and following up on proposals ensures successful development and deployment of BIS technology.

Ethical and Societal Impact of Bullet Inversion Systems (BIS)

The development and deployment of advanced technologies like the Bullet Inversion System (BIS) carry significant ethical and societal implications. While such technologies offer the potential for enhanced protection and safety, they also raise complex questions about their broader impact on society and the ethical considerations surrounding their use. This discussion explores the societal impact of BIS, including its potential effects on public safety, privacy, and social dynamics, as well as the ethical issues related to its deployment and use.

Assessing the Broader Societal Impact of BIS

1. Enhancing Public Safety

The primary societal benefit of BIS technology lies in its potential to significantly enhance public safety. By providing a protective mechanism capable of deflecting or neutralizing bullets, BIS could reduce injuries and fatalities resulting from firearms. This improvement in personal protection could lead to a decrease in crime rates, particularly in high-risk environments such as conflict zones or areas with elevated gun violence.

1.1. Reducing Casualties

In contexts where BIS is deployed, such as in military or law enforcement operations, the technology could substantially reduce casualties among personnel. The ability to neutralize or deflect bullets could protect individuals from harm during active engagements, thereby potentially saving lives and reducing the need for medical intervention.

1.2. Impact on Emergency Services

For emergency services and first responders, BIS technology could offer added protection during critical interventions. By reducing the risk of injury from gunfire, BIS could improve the safety of personnel involved in high-risk situations, such as hostage rescues or active shooter scenarios.

2. Privacy and Civil Liberties

While the safety benefits of BIS are significant, the technology's deployment raises concerns about privacy and civil liberties. The integration of BIS into public spaces or personal protective gear could lead to increased surveillance and monitoring.

2.1. Surveillance Concerns

Incorporating BIS technology into public safety measures may involve the use of advanced sensors and monitoring systems. This could raise privacy concerns, as the collection and analysis of data related to gunfire and threats might lead to unintended surveillance of individuals. Balancing the benefits of enhanced protection with the need to protect personal privacy is a crucial consideration.

2.2. Civil Liberties

The use of BIS in public spaces or private properties could impact civil liberties, particularly if the technology is used to enforce stricter security measures. Ensuring that BIS deployment does not infringe on individuals' rights or lead to unnecessary restrictions on personal freedoms is an important aspect of its ethical implementation.

3. Social Dynamics and Equity

The introduction of BIS technology could influence social dynamics and raise questions of equity and access. Ensuring that the technology is deployed in a manner that promotes social fairness and addresses disparities is essential.

3.1. Access and Inequality

The cost of BIS technology could create disparities in access, with higher protection levels available only to those who can afford it. This potential inequality could lead to a scenario where only certain segments of society benefit from advanced protective measures, while others remain vulnerable. Addressing issues of access and affordability is critical to prevent exacerbating social inequalities.

3.2. Impact on Crime and Violence

The deployment of BIS technology could also impact patterns of crime and violence. While the technology may deter certain types of violent incidents, it could potentially lead to unintended consequences, such as shifts in criminal behavior or the development of counter-technologies. Understanding and mitigating these potential effects is important for ensuring that BIS contributes positively to societal safety.

Ethical Considerations in Deployment and Use

1. Responsible Development and Implementation

The ethical development and implementation of BIS technology require careful consideration of various factors to ensure that the technology is used responsibly and aligns with societal values.

1.1. Transparency and Accountability

Transparency in the development and deployment of BIS is essential for maintaining public trust. This includes clear communication about the technology's capabilities, limitations, and intended use. Accountability measures should be established to address any misuse or unintended consequences, ensuring that the technology is used ethically and in accordance with established guidelines.

1.2. Informed Consent

In contexts where BIS technology is deployed, particularly in public spaces or as part of personal protective equipment, obtaining informed consent from individuals affected by its use is crucial. Ensuring that individuals are aware of and understand how the technology will be used and its potential implications is a key ethical consideration.

2. Ethical Use and Misuse

Ensuring that BIS technology is used ethically involves addressing concerns related to its potential misuse and the broader implications of its application.

2.1. Proportionality and Necessity

The principle of proportionality requires that BIS technology be used only when necessary and appropriate for the intended purpose. Deploying BIS in situations where less intrusive measures would suffice could be considered ethically problematic. Ensuring that the technology is used in a manner that is proportionate to the threat and necessary for achieving the desired outcomes is important for maintaining ethical standards.

2.2. Potential for Abuse

The potential for abuse of BIS technology must be carefully managed. This includes preventing its use in ways that could infringe on individual rights or contribute to authoritarian practices. Establishing safeguards and oversight mechanisms to prevent misuse is essential for upholding ethical standards.

3. Societal and Cultural Impact

The broader societal and cultural impact of BIS technology must be considered, including its potential effects on societal norms and values.

3.1. Cultural Sensitivity

Cultural sensitivity is important when deploying BIS technology in diverse settings. Different cultures may have varying attitudes toward personal protection and security technologies. Understanding and respecting these cultural perspectives is crucial for ensuring that the technology is implemented in a manner that aligns with local values and practices.

3.2. Public Perception and Trust

The public perception of BIS technology will influence its acceptance and effectiveness. Building public trust through transparent communication, ethical practices, and demonstrated benefits is key to gaining societal support. Engaging with communities, addressing concerns, and providing education about the technology can help foster positive public perception.

Conclusion

The ethical and societal impact of Bullet Inversion Systems (BIS) encompasses a range of considerations, from enhancing public safety to addressing privacy, equity, and cultural factors. While BIS technology offers significant potential benefits, its deployment must be managed responsibly to avoid negative consequences and ensure alignment with ethical principles. Balancing the advantages of improved protection with the need to safeguard privacy, equity, and societal values is essential for the responsible development and implementation of BIS technology. Engaging in thoughtful analysis and open dialogue about these issues will contribute to the ethical advancement of BIS and its positive integration into society.

Intellectual Property and Patents for Bullet Inversion Systems (BIS)

Intellectual property (IP) and patent strategies play a crucial role in the development, protection, and commercialization of advanced technologies like the Bullet Inversion System (BIS). The BIS technology, which represents a significant innovation in personal protection and defense mechanisms, necessitates careful consideration of IP laws and strategic patent management to safeguard innovations and gain competitive advantages. This discussion delves into navigating the intellectual property landscape and devising effective patent strategies for BIS technology.

Navigating the Intellectual Property Landscape

1. Understanding Intellectual Property Rights

Intellectual property encompasses various rights that protect innovations and creations from unauthorized use. For BIS technology, several types of IP rights are relevant:

1.1. Patents
Patents grant exclusive rights to inventors for a limited time, usually 20 years, in exchange for public disclosure of their invention. They cover novel and non-obvious inventions, including mechanisms, systems, and processes integral to BIS. Securing patents ensures that competitors cannot replicate or exploit the technology without permission.

1.2. Trademarks
Trademarks protect brands, logos, and names associated with BIS technology. They help in distinguishing the BIS products and services from competitors in the market. Registering trademarks can enhance brand recognition and protect the technology's market identity.

1.3. Trade Secrets
Trade secrets involve confidential business information that provides a competitive edge. For BIS technology, this might include proprietary designs, manufacturing processes, or algorithmic methods that are not publicly disclosed but are critical to the technology's functionality and performance.

1.4. Copyrights
Copyrights protect original works of authorship, such as technical documentation, software code, and marketing materials associated with BIS. Although not directly applicable to the hardware itself, copyrights can safeguard written and digital content related to BIS technology.

2. Conducting IP Research

Effective IP management begins with comprehensive research to understand existing patents and potential IP conflicts. Key steps include:

2.1. Patent Searches
Performing detailed patent searches helps identify existing patents that may overlap with BIS technology. This includes searching through patent databases to determine whether similar inventions have been patented and assessing their relevance to the BIS.

2.2. Freedom-to-Operate Analysis
A freedom-to-operate (FTO) analysis assesses the risk of infringing on existing patents. By

analyzing current patents and pending applications, companies can ensure that their BIS technology does not violate any IP rights and can proceed with development and commercialization with reduced legal risk.

2.3. IP Landscaping

IP landscaping involves mapping out the IP environment related to BIS technology. This includes identifying key players, understanding trends, and recognizing gaps in the market that BIS technology might fill. IP landscaping helps in strategic planning and identifying potential areas for innovation.

Patent Strategies for BIS Technology

1. Developing a Patent Portfolio

Building a robust patent portfolio is essential for protecting BIS technology and gaining competitive advantages. Effective patent strategies include:

1.1. Filing Broad and Specific Patents

Filing patents that cover both broad and specific aspects of BIS technology ensures comprehensive protection. Broad patents might cover general principles and overall systems, while specific patents focus on detailed components, mechanisms, or applications. This approach helps safeguard the core innovations and any ancillary developments.

1.2. International Patenting

Given the global nature of technology markets, securing international patents is crucial for BIS technology. Filing patents in key markets ensures that the technology is protected worldwide and prevents competitors from exploiting the technology in different jurisdictions. The Patent Cooperation Treaty (PCT) offers a streamlined process for international patent applications.

1.3. Patent Continuations and Divisions

Utilizing patent continuations and divisions allows for the refinement and expansion of patent claims. Continuations enable additional claims to be added based on the original patent application, while divisions split a single application into multiple applications to cover different aspects of the technology.

2. Managing Patent Lifecycles

Effectively managing the lifecycle of patents involves:

2.1. Monitoring Patent Expiration

Keeping track of patent expiration dates is essential for planning future innovations and protecting intellectual property. Once a patent expires, the technology enters the public domain, so timely renewal or new filings are necessary to maintain protection.

2.2. Enforcing Patent Rights

Enforcing patent rights involves monitoring competitors for potential infringements and taking legal action when necessary. Establishing an IP enforcement strategy ensures that any unauthorized use of BIS technology is addressed promptly and effectively.

3. Licensing and Collaboration

3.1. Licensing Agreements
Licensing BIS technology to other companies or entities can generate revenue and expand market reach. Licensing agreements should be carefully negotiated to ensure fair terms and protect IP rights while enabling commercial partnerships.

3.2. Collaborative Research
Collaborating with research institutions or other companies can enhance BIS technology development and access additional expertise. Joint research agreements should include clear terms regarding IP ownership, contribution, and commercialization rights.

4. IP Strategy Alignment

Aligning IP strategy with overall business goals is crucial for maximizing the value of BIS technology:

4.1. Strategic Alignment
Ensure that IP strategies align with business objectives, such as market entry, competitive positioning, and product development. This alignment helps in prioritizing IP efforts and resources to support business growth and innovation.

4.2. IP Valuation
Regularly assessing the value of IP assets helps in making informed decisions about investment, licensing, and commercialization. IP valuation considers factors such as market potential, competitive advantage, and technological significance.

5. Addressing IP Risks

Managing IP risks involves:

5.1. IP Litigation Risks
Understanding and preparing for potential IP litigation risks is essential. This includes assessing the likelihood of infringement claims and developing strategies to mitigate legal disputes.

5.2. IP Theft and Counterfeiting
Protecting BIS technology from theft and counterfeiting involves implementing security measures, monitoring markets, and enforcing IP rights. Addressing these risks helps maintain the integrity of the technology and its commercial value.

Conclusion

Navigating the intellectual property landscape and developing effective patent strategies are vital for the success of Bullet Inversion Systems (BIS) technology. By understanding the various types of IP rights, conducting thorough research, and implementing strategic patent management practices, companies can protect their innovations, gain competitive advantages, and ensure successful commercialization. Addressing IP risks and aligning IP strategies with business goals further enhances the value and impact of BIS technology. As BIS technology continues to evolve, ongoing attention to intellectual property considerations will be essential for sustaining innovation and achieving long-term success.

Marketing and Commercialization Strategies for Bullet Inversion Systems (BIS)

The development and successful deployment of advanced technologies like the Bullet Inversion System (BIS) hinge significantly on effective marketing and commercialization strategies. As a groundbreaking innovation in personal protection and defense, BIS requires a well-structured approach to market entry, customer engagement, and sustainable growth. This discussion explores the essential components of developing a marketing plan for BIS, identifying target markets, and outlining commercialization pathways.

Developing a Marketing Plan for BIS

1. Defining the Value Proposition

1.1. Unique Selling Points (USPs)
 The value proposition of BIS must clearly articulate its unique selling points. These might include its advanced bullet deflection capabilities, integration with existing protective technologies, or innovative design features that set it apart from traditional body armor. Emphasizing how BIS addresses specific pain points, such as enhanced protection and real-time threat response, can effectively communicate its value to potential customers.

1.2. Benefits and Features
 Highlighting the tangible benefits of BIS, such as improved safety and operational efficiency, is crucial. This involves detailing features like its advanced magnetic field generation or plasma shield technology. Demonstrating how these features translate into practical advantages for users, such as reduced risk of injury or enhanced situational awareness, helps in building a compelling marketing message.

2. Market Research and Analysis

2.1. Industry Trends and Market Needs
 Conducting comprehensive market research helps in understanding current industry trends, emerging technologies, and market needs. Identifying gaps in the market where BIS can provide significant improvements or advantages over existing solutions is key. Research should focus on the evolving landscape of personal protection technologies and the demand for advanced defense systems.

2.2. Competitive Analysis

Analyzing competitors, including their product offerings, market positioning, and customer feedback, provides insights into BIS's competitive landscape. Understanding competitors' strengths and weaknesses allows for identifying unique opportunities for BIS and differentiating it effectively. This analysis should include a review of both direct competitors (other personal protection systems) and indirect competitors (traditional body armor and security technologies).

3. Developing a Marketing Strategy

3.1. Target Audience Identification

Identifying the target audience for BIS involves segmenting the market based on various criteria such as industry, geographic location, and specific needs. Potential target markets for BIS include military and defense organizations, law enforcement agencies, and high-security private entities. Each segment may have unique requirements and preferences, necessitating tailored marketing approaches.

3.2. Branding and Messaging

Developing a strong brand identity for BIS is essential for market recognition and credibility. This includes creating a compelling brand name, logo, and tagline that resonate with the target audience. Consistent messaging across all marketing channels reinforces the BIS brand and helps in establishing a positive reputation.

3.3. Marketing Channels and Tactics

Selecting the appropriate marketing channels and tactics is critical for reaching the target audience effectively. This may include:

- **Digital Marketing**: Utilizing online platforms such as social media, search engine optimization (SEO), and content marketing to increase visibility and engage potential customers.

- **Trade Shows and Conferences**: Participating in industry events to showcase BIS technology, network with key stakeholders, and generate leads.

- **Direct Marketing**: Implementing targeted email campaigns, webinars, and personalized outreach to engage with specific prospects.

- **Public Relations**: Leveraging media coverage, press releases, and thought leadership articles to build credibility and raise awareness about BIS.

4. Sales and Distribution Channels

4.1. Direct Sales

Establishing a direct sales force allows for personalized interactions with potential customers and a deeper understanding of their needs. Direct sales can include demonstrations, consultations, and tailored solutions for specific applications of BIS.

4.2. Distribution Partnerships
Partnering with established distributors and resellers in the defense and security sectors can accelerate market penetration. These partnerships provide access to established networks and customer bases, facilitating broader reach and faster adoption of BIS technology.

4.3. Online Sales
Creating an online platform or e-commerce site for BIS enables direct sales and provides detailed product information, specifications, and purchasing options. An online presence also supports global reach and accessibility for customers in various regions.

Identifying Target Markets and Commercialization Pathways

1. Market Segmentation

1.1. Military and Defense
The military and defense sector is a primary target market for BIS technology due to its focus on advanced protection solutions. BIS can offer significant enhancements over traditional body armor, particularly in high-risk combat scenarios. Engaging with military procurement agencies and defense contractors is essential for understanding specific requirements and integrating BIS into defense systems.

1.2. Law Enforcement Agencies
Law enforcement agencies require advanced protective gear for officers operating in high-threat environments. BIS can enhance officer safety and operational effectiveness. Collaborating with police departments and security forces to demonstrate BIS technology and gather feedback can facilitate adoption and customization.

1.3. Private Security and High-Net-Worth Individuals
The private security sector, including high-net-worth individuals and corporate security teams, represents another potential market. BIS can offer added protection for executive security and high-profile events. Tailoring marketing efforts to address the unique needs of private security and offering bespoke solutions can drive interest and sales.

2. Commercialization Pathways

2.1. Strategic Partnerships and Alliances
Forming strategic partnerships with technology developers, research institutions, and industry leaders can accelerate the commercialization of BIS. Collaborative efforts can include joint development projects, co-marketing initiatives, and shared expertise. These partnerships can also facilitate access to additional resources and funding.

2.2. Licensing Agreements
Licensing BIS technology to other companies or entities can generate revenue and expand market reach. Licensing agreements should clearly outline terms related to technology use, royalties, and intellectual property rights. This approach allows for leveraging external expertise and resources while maintaining control over core technology.

2.3. Government Contracts and Tenders

Securing government contracts and tenders is a viable pathway for commercialization, particularly in defense and security sectors. Engaging in government procurement processes and responding to tenders can lead to significant opportunities for BIS technology deployment and funding.

2.4. Pilot Projects and Trials

Conducting pilot projects and trials with select customers or partners provides an opportunity to validate BIS technology in real-world conditions. Successful pilot programs can generate valuable case studies and endorsements, which can be leveraged for broader market adoption and commercialization.

Conclusion

Developing a comprehensive marketing plan and identifying effective commercialization pathways are essential for the successful deployment of Bullet Inversion Systems (BIS) technology. By defining a clear value proposition, conducting thorough market research, and implementing targeted marketing strategies, companies can effectively engage with potential customers and drive adoption. Identifying key target markets and exploring diverse commercialization pathways, such as strategic partnerships, licensing agreements, and government contracts, further enhances the potential for BIS technology to achieve widespread impact and success.

Feedback and Iterative Design for Bullet Inversion Systems (BIS)

The development of advanced technologies like Bullet Inversion Systems (BIS) necessitates a robust approach to feedback and iterative design. These methodologies are critical in refining technology, ensuring its effectiveness, and enhancing user satisfaction. This discussion explores the significance of user feedback in the refinement of BIS, and the iterative design process essential for continuous improvement.

Importance of User Feedback in Refining BIS

1. Understanding User Needs and Expectations

1.1. Gathering Insights from End-Users

User feedback provides valuable insights into the practical needs and expectations of those who will interact with BIS technology. Engaging with potential users, such as military personnel, law enforcement officers, and security professionals, allows for the identification of specific requirements and potential areas for improvement. This feedback can highlight usability issues, performance concerns, and desired features that may not be apparent during the initial development phase.

1.2. Identifying Pain Points and Areas for Improvement

Real-world use often uncovers issues that are not detected during laboratory testing or simulations. Feedback from users can reveal pain points related to BIS's functionality, comfort, or integration with other equipment. Addressing these concerns promptly is crucial for enhancing the overall performance and acceptance of the system.

2. Enhancing Performance and Effectiveness

2.1. Fine-Tuning System Performance
User feedback helps in fine-tuning the performance of BIS by providing data on how well the system functions in diverse scenarios. For example, feedback on the effectiveness of bullet deflection, the reliability of magnetic fields, or the stability of plasma shields allows engineers to make precise adjustments and improvements. This iterative process ensures that BIS performs optimally in the field.

2.2. Validating Design Choices
Design choices made during the development of BIS, such as material selection, system configuration, and integration methods, need validation through real-world use. User feedback serves as a critical check to confirm that these design choices align with practical needs and expectations. Continuous validation ensures that BIS technology remains relevant and effective.

3. Building User Trust and Satisfaction

3.1. Engaging Users in the Development Process
Involving users in the development process fosters trust and satisfaction. By actively seeking and incorporating feedback, developers demonstrate a commitment to addressing user needs and enhancing the system. This engagement builds confidence in BIS technology and encourages user buy-in.

3.2. Improving Usability and Acceptance
User feedback is instrumental in improving the usability of BIS. Feedback on user interfaces, system controls, and overall user experience helps in designing a more intuitive and user-friendly system. Enhanced usability leads to higher acceptance and more effective use of BIS technology in practical situations.

Iterative Design Process for Continuous Improvement

1. Theoretical Foundation of Iterative Design

1.1. Concept of Iterative Design
Iterative design is a methodology that involves repeating cycles of design, testing, and refinement to gradually improve a product. This approach is based on the principle that continuous feedback and incremental changes lead to a more refined and effective end product. In the context of BIS, iterative design allows for ongoing enhancements and adaptation to evolving needs.

1.2. Application in Technology Development
In technology development, iterative design involves several stages, including concept

development, prototype testing, and system refinement. Each iteration provides opportunities to address issues, incorporate feedback, and optimize performance. This cyclical process ensures that BIS evolves in response to real-world conditions and user needs.

2. Practical Implementation of Iterative Design

2.1. Prototype Development and Testing
The iterative design process begins with the creation of prototypes that embody the initial design concepts. These prototypes are tested under controlled conditions to evaluate their performance and gather user feedback. Testing helps identify strengths and weaknesses, guiding subsequent iterations and improvements.

2.2. Incorporating Feedback into Design
Feedback from prototype testing is analyzed to inform design changes. Engineers and designers review user comments, performance data, and observations to identify areas for enhancement. Changes are made to address issues and improve the system's functionality, usability, and overall effectiveness.

2.3. Repeating the Cycle
The refined prototypes undergo additional rounds of testing and feedback collection. Each iteration involves implementing changes based on previous feedback, followed by further testing to validate improvements. This cycle continues until the BIS meets the desired performance criteria and user satisfaction levels.

3. Logical Considerations for Iterative Design

3.1. Continuous Improvement and Adaptation
Iterative design ensures that BIS technology is continuously improved and adapted to changing needs. By incorporating feedback and refining designs, developers can address emerging challenges and leverage new opportunities. This approach promotes ongoing innovation and relevance in a rapidly evolving technological landscape.

3.2. Managing Resources and Time
While iterative design offers numerous benefits, it requires effective management of resources and time. Each iteration involves costs related to prototype development, testing, and analysis. Balancing these factors with the need for continuous improvement is essential for maintaining efficiency and ensuring timely advancements.

4. Expansive View of Iterative Design in Technology

4.1. Impact on Technology Evolution
Iterative design plays a crucial role in the evolution of technology by fostering a culture of continuous improvement and user-centered development. It allows for the gradual refinement of complex systems, such as BIS, and facilitates the incorporation of new insights, technologies, and innovations.

4.2. Collaboration and Feedback Integration

Successful iterative design involves collaboration among various stakeholders, including developers, users, and industry experts. Integrating diverse perspectives and expertise enriches the feedback process and enhances the overall quality of the BIS technology. This collaborative approach also ensures that the system meets the diverse needs of its users.

4.3. Future Directions and Innovations

The iterative design process opens avenues for exploring new technologies and approaches. As BIS technology advances, iterative design will continue to drive innovation by incorporating emerging trends, such as advanced materials, novel power sources, and improved safety features. This dynamic process ensures that BIS remains at the forefront of personal protection technology.

Conclusion

The feedback and iterative design process is fundamental to the successful development and refinement of Bullet Inversion Systems (BIS). User feedback provides essential insights into system performance and usability, guiding improvements and ensuring that BIS meets practical needs. The iterative design process, characterized by cycles of testing, refinement, and adaptation, fosters continuous improvement and innovation. By integrating feedback effectively and embracing iterative methodologies, developers can enhance BIS technology, build user trust, and achieve greater success in the market.

Workshops and Conferences: Sharing Research and Networking in the Advancement of Technologies

1. Organizing Workshops and Conferences

1.1. Objectives and Planning

1.1.1. Defining Goals and Scope

Organizing workshops and conferences is a pivotal strategy for disseminating research, fostering collaboration, and advancing technology. The primary objective is to create a platform where researchers, practitioners, and stakeholders can exchange knowledge, discuss developments, and identify future directions. Planning begins with defining clear goals and the scope of the event, tailored to the specific needs and interests of the target audience. This involves setting objectives such as showcasing recent research, discussing emerging trends, or solving industry-specific challenges.

1.1.2. Structuring the Program

The structure of the event should be designed to facilitate comprehensive discussions and interactions. A typical agenda includes keynote speeches from leading experts, panel discussions, technical sessions, and breakout groups. Keynote addresses offer insights into major advancements and trends, while technical sessions and panels provide in-depth analysis and discussions on specialized topics. Interactive breakout groups and workshops allow for hands-on learning and problem-solving, encouraging active participation and engagement from attendees.

1.1.3. Selecting Speakers and Experts

Inviting prominent speakers and experts is crucial for ensuring the event's success. These individuals bring credibility, expertise, and a broad perspective to the discussions. When selecting speakers, consider their contributions to the field, their ability to communicate complex ideas effectively, and their relevance to the event's themes. Engaging a diverse range of experts from academia, industry, and government can provide a well-rounded view of current trends and future directions.

1.2. Logistics and Execution

1.2.1. Venue and Technology

Choosing an appropriate venue is essential for accommodating attendees and ensuring the smooth execution of the event. The venue should be equipped with necessary facilities, such as audio-visual equipment, internet access, and comfortable seating. Modern technology, including virtual conferencing tools, can expand the reach of the event and facilitate remote participation. Ensuring seamless integration of technology is key to providing a professional and engaging experience for all attendees.

1.2.2. Registration and Promotion

Effective registration and promotion strategies are vital for attracting participants and managing attendance. Implementing an online registration system simplifies the process for attendees and organizers. Promotion through academic journals, industry newsletters, social media, and professional networks helps raise awareness and encourages participation. Early promotion and reminders can boost attendance and engagement.

1.2.3. Post-Event Evaluation

After the event, conducting evaluations and collecting feedback from attendees is important for assessing the success of the workshop or conference. Surveys and feedback forms can provide insights into what worked well and what areas need improvement. Analyzing this feedback helps in refining future events and ensuring continuous improvement in the quality of presentations and interactions.

2. Networking with Experts and Stakeholders

2.1. Importance of Networking

2.1.1. Building Professional Relationships

Networking at workshops and conferences offers opportunities to build professional relationships with experts, stakeholders, and peers. These connections can lead to collaborations, joint research projects, and partnerships that advance technology and innovation. Establishing a network of contacts within the field fosters a collaborative environment and facilitates knowledge exchange.

2.1.2. Sharing Knowledge and Insights
Networking events provide a platform for sharing insights, discussing challenges, and exploring new ideas. Informal discussions and interactions during breaks, social events, and networking sessions enable participants to exchange knowledge and gain different perspectives. This informal knowledge sharing can complement formal presentations and contribute to a more comprehensive understanding of the subject matter.

2.2. Strategies for Effective Networking

2.2.1. Preparation and Goals
Effective networking begins with preparation. Attendees should set specific goals for the event, such as meeting potential collaborators, exploring new technologies, or seeking feedback on their work. Researching key participants and speakers beforehand can help in identifying relevant individuals to connect with and prepare targeted questions or discussion points.

2.2.2. Engaging in Meaningful Conversations
Initiating and engaging in meaningful conversations is crucial for successful networking. Participants should approach discussions with an open mind, actively listen, and contribute valuable insights. Building rapport through genuine interest and thoughtful dialogue enhances the quality of interactions and fosters stronger professional relationships.

2.2.3. Leveraging Networking Tools
Utilizing networking tools, such as event apps and social media platforms, can enhance connectivity and engagement. Many conferences offer dedicated apps for scheduling meetings, accessing session materials, and connecting with other attendees. Social media platforms, such as LinkedIn and Twitter, provide additional avenues for maintaining connections and continuing conversations beyond the event.

2.3. Following Up and Maintaining Connections

2.3.1. Post-Event Follow-Up
Following up with contacts made during the event is essential for maintaining and strengthening relationships. Sending personalized follow-up emails or messages helps to reinforce connections and express appreciation for the conversations. Sharing relevant information, such as research findings or articles, can keep the dialogue active and provide additional value.

2.3.2. Building Long-Term Relationships
Networking is not a one-time activity but an ongoing process. Maintaining connections through regular communication, collaboration, and participation in future events helps to build long-term professional relationships. Engaging with contacts on social media, attending related events, and contributing to collaborative projects can further solidify these connections.

3. Theoretical and Practical Implications

3.1. Theoretical Foundations of Networking
The theoretical framework of networking emphasizes the importance of social capital and

relationship-building in advancing professional and academic goals. Networking theories, such as social network theory and the strength of weak ties, highlight how connections and interactions contribute to the flow of information and opportunities. Understanding these theories can guide effective networking strategies and enhance the impact of workshops and conferences.

3.2. Practical Considerations for Organizers and Participants

For organizers, practical considerations include ensuring smooth event execution, facilitating effective interactions, and providing resources for networking. For participants, practical strategies involve setting clear goals, engaging actively, and leveraging available tools. Balancing theoretical insights with practical applications enhances the overall success of workshops and conferences.

4. Expansive View of Workshops and Conferences

4.1. Impact on Research and Development

Workshops and conferences play a crucial role in driving research and development by facilitating knowledge exchange, fostering collaborations, and stimulating innovation. They serve as platforms for showcasing new research, discussing emerging trends, and identifying research gaps. The insights and connections gained from these events contribute to the advancement of technology and scientific understanding.

4.2. Future Directions and Innovations

The future of workshops and conferences will likely involve increased integration of digital technologies, such as virtual and hybrid events, to expand reach and accessibility. Innovations in event formats, such as interactive sessions and real-time feedback mechanisms, will further enhance the participant experience. Embracing these advancements will continue to drive progress and collaboration in various fields.

Conclusion

Organizing workshops and conferences is a vital strategy for advancing research, fostering collaboration, and driving technological innovation. Effective planning, execution, and networking are key to the success of these events. By understanding the importance of user feedback, implementing iterative design processes, and leveraging networking opportunities, stakeholders can maximize the impact of workshops and conferences. Embracing theoretical insights and practical strategies ensures that these events contribute significantly to the advancement of technology and knowledge.

Educational Outreach and Awareness: Advancing Knowledge and Engagement on BIS Technology

1. Creating Educational Programs to Raise Awareness about BIS

1.1. Objectives and Rationale

1.1.1. Defining Educational Goals
Educational outreach is essential for raising awareness and understanding of emerging technologies like the Bullet Inversion System (BIS). The primary objective is to inform various audiences about the benefits, workings, and implications of BIS technology. This includes creating educational programs that are informative, engaging, and accessible, catering to different knowledge levels, from general public to specialized professionals. These programs aim to foster a well-informed community that can critically assess and contribute to the development and application of BIS technology.

1.1.2. Target Audience Identification
Identifying the target audience is crucial for tailoring educational programs effectively. Potential audiences include students, educators, industry professionals, policymakers, and the general public. Each group has different informational needs and levels of understanding, necessitating customized content and delivery methods. For instance, high school students might benefit from interactive demonstrations and simplified explanations, while industry professionals may require in-depth technical seminars and case studies.

1.2. Program Development and Implementation

1.2.1. Curriculum Design and Content Creation
Developing a curriculum involves creating content that accurately represents the principles and applications of BIS technology. This includes designing modules that cover fundamental concepts such as electromagnetic principles, plasma physics, and the practical applications of BIS. Interactive elements like simulations, hands-on experiments, and visual aids can enhance understanding. Content should be engaging and updated with the latest advancements to maintain relevance and accuracy.

1.2.2. Educational Materials and Resources
Providing a variety of educational materials and resources is key to reaching different learning styles. This may include textbooks, online resources, instructional videos, and interactive tools. Digital platforms can be leveraged to create accessible and scalable learning experiences, such as e-learning modules, webinars, and virtual workshops. Collaborations with educational technology developers can enhance these resources and make learning more engaging.

1.2.3. Professional Development for Educators
Training educators to effectively deliver educational content on BIS is crucial. This involves providing professional development workshops and resources that equip educators with the knowledge and skills to teach the material. Support materials like lesson plans, teaching guides, and assessment tools can help educators integrate BIS topics into their existing curricula. Ongoing support and feedback mechanisms can ensure that educators remain up-to-date with advancements in the field.

2. Engaging with Schools and Community Organizations

2.1. Partnerships and Collaborations

2.1.1. Building Relationships with Schools

Establishing partnerships with schools is a strategic approach to integrating BIS education into K-12 curricula. Collaboration with school administrators and educators can help design and implement programs that align with educational standards and objectives. Activities might include guest lectures, workshops, and interactive demonstrations. Schools can also host science fairs and exhibitions showcasing BIS technology, fostering student interest and engagement.

2.1.2. Community Organization Engagement

Community organizations, such as local science centers, libraries, and civic groups, offer additional platforms for educational outreach. Partnering with these organizations can expand the reach of BIS education and engage diverse audiences. Programs might include public lectures, community workshops, and informational events. Collaboration with non-profit organizations focused on science and technology can enhance program visibility and impact.

2.2. Outreach Strategies and Activities

2.2.1. Interactive Demonstrations and Workshops

Hands-on activities and interactive demonstrations are effective in illustrating complex concepts related to BIS. Workshops that involve building simple electromagnetic devices or simulations of BIS principles can provide practical experience and enhance understanding. These activities encourage active learning and allow participants to explore concepts in a tangible way.

2.2.2. Educational Events and Exhibitions

Organizing educational events and exhibitions can increase public awareness and interest in BIS technology. Science fairs, technology expos, and public lectures are opportunities to showcase BIS applications and advancements. Engaging presentations, live demonstrations, and interactive exhibits can capture the audience's attention and provide a deeper understanding of BIS technology.

2.2.3. Digital Outreach and Social Media

Leveraging digital platforms and social media is crucial for reaching a wider audience. Creating engaging content such as blogs, videos, and infographics about BIS technology can inform and educate the public. Social media campaigns and online webinars can facilitate real-time interaction and discussions, making it easier for individuals to learn about BIS and its applications. Online forums and discussion groups can also provide platforms for ongoing engagement and knowledge sharing.

2.3. Evaluating and Enhancing Outreach Efforts

2.3.1. Measuring Program Effectiveness

Evaluating the effectiveness of educational programs is essential for continuous improvement. Surveys, feedback forms, and assessments can provide insights into participant learning outcomes and program impact. Analyzing this data helps identify strengths and areas for improvement, allowing for adjustments to enhance program effectiveness and engagement.

2.3.2. Iterative Improvement and Adaptation

Based on evaluation results, programs should be regularly updated and refined. Incorporating feedback from participants and educators can help in making necessary adjustments and ensuring that content remains relevant and engaging. Adapting programs to reflect new developments in BIS technology and changes in educational standards ensures ongoing relevance and effectiveness.

3. Theoretical and Practical Implications

3.1. Theoretical Foundations of Educational Outreach

Educational outreach is grounded in theories of learning and communication. Constructivist theories emphasize active learning and the construction of knowledge through experience and interaction. Theoretical frameworks such as experiential learning and social learning theory highlight the importance of engaging and participatory methods in education. Applying these theories to BIS outreach ensures that educational programs are effective in fostering understanding and interest.

3.2. Practical Implementation Strategies

Practical implementation involves designing and executing programs that align with theoretical principles. This includes developing engaging content, using diverse educational materials, and leveraging multiple platforms for outreach. Collaboration with educators, community organizations, and technology developers enhances the reach and impact of educational programs.

4. Expansive View of Educational Outreach

4.1. Long-Term Impact on Technology Adoption

Educational outreach plays a crucial role in shaping public perception and adoption of emerging technologies like BIS. By raising awareness and understanding, educational programs can influence attitudes towards technology and its applications. Increased knowledge can lead to greater acceptance, informed decision-making, and support for further research and development.

4.2. Future Directions and Innovations

The future of educational outreach will likely involve advancements in digital technologies and interactive learning tools. Innovations such as virtual reality (VR) and augmented reality (AR) can provide immersive learning experiences, making complex concepts more accessible. Embracing these technologies will enhance the effectiveness and reach of educational programs, ensuring that they remain relevant and impactful.

Conclusion

Educational outreach and awareness are fundamental to advancing understanding and acceptance of BIS technology. Creating comprehensive educational programs, engaging with schools and community organizations, and leveraging digital platforms are key strategies for effective outreach. By applying theoretical principles and practical implementation strategies, stakeholders can enhance public knowledge, foster interest, and contribute to the successful integration of BIS technology into society. The ongoing evolution of educational methods and technologies will continue to shape the future of outreach, driving innovation and engagement in the field.

Technological Readiness Levels (TRL): Evaluating and Advancing the Bullet Inversion System (BIS)

1. Understanding Technological Readiness Levels (TRL)

1.1. Definition and Purpose

Technological Readiness Levels (TRLs) are a systematic metric used to evaluate the maturity of a technology. Developed by NASA, TRLs provide a framework for assessing the progress of a technology from initial concept through to full deployment. The TRL scale ranges from 1 to 9, with each level representing a stage in the development lifecycle. The purpose of TRLs is to offer a standardized method for gauging technological maturity, identifying development needs, and managing risk throughout the technology's development.

1.2. TRL Framework

1.2.1. TRL 1-3: Conceptualization and Feasibility

- **TRL 1:** Basic principles observed. At this stage, fundamental scientific principles related to BIS are identified and explored.

- **TRL 2:** Technology concept formulated. Initial theoretical models for BIS are developed, focusing on core concepts like electromagnetic fields and plasma generation.

- **TRL 3:** Analytical and experimental proof of concept. Early experiments demonstrate the feasibility of the BIS concepts in a controlled laboratory setting.

1.2.2. TRL 4-6: Development and Testing

- **TRL 4:** Component and/or breadboard validation in a laboratory environment. At this stage, small-scale prototypes of BIS components are tested under controlled conditions to validate design and functionality.

- **TRL 5:** Component and/or breadboard validation in a relevant environment. The technology is tested in a simulated or operational environment that closely resembles real-world conditions.

- **TRL 6:** System/subsystem model or prototype demonstration in a relevant environment. A more integrated prototype of BIS is demonstrated, showing how different components work together in a relevant operational environment.

1.2.3. TRL 7-9: System Integration and Deployment

- **TRL 7:** System prototype demonstration in an operational environment. The BIS prototype is tested in real-world conditions, demonstrating its performance and reliability.

- **TRL 8:** Actual system completed and qualified through test and demonstration. The BIS system undergoes rigorous testing and validation to ensure it meets all operational requirements.

- **TRL 9:** Actual system proven through successful mission operations. BIS technology is fully operational, integrated, and demonstrated in its intended operational environment.

2. Assessing BIS Against TRL Metrics

2.1. Current TRL Assessment

2.1.1. Evaluating BIS Development Stages
To assess BIS against TRL metrics, we need to examine its current stage in development. For instance, if BIS is in the early stages of concept development, it may be at TRL 1 or 2. The focus is on theoretical models and preliminary experiments. As the technology advances, it progresses through TRL 3 with experimental proof of concept and eventually moves towards higher TRLs as prototypes are developed and tested.

2.1.2. Identifying Gaps and Requirements
Identifying gaps between the current TRL and target TRLs is crucial for planning the next development stages. This involves assessing the readiness of individual components, system integration, and testing procedures. For example, if BIS is currently at TRL 4, the next steps involve validating components in relevant environments (TRL 5) and demonstrating an integrated prototype (TRL 6).

2.2. Developing a Roadmap to Higher TRLs

2.2.1. Setting Milestones and Objectives
To advance BIS to higher TRLs, clear milestones and objectives must be established. These milestones should include specific goals for component development, system integration, and testing. Each milestone should have defined criteria for success, such as achieving specific performance metrics or completing successful demonstrations in relevant environments.

2.2.2. Resource Allocation and Planning
Effective planning requires allocating resources, including funding, personnel, and facilities. Resource allocation involves identifying the required budget for development, securing necessary equipment and materials, and assembling a team with the expertise needed for advancing BIS. Planning should also include timelines for achieving each milestone and contingencies for potential challenges.

2.2.3. Risk Management and Mitigation

Managing and mitigating risks is essential for advancing BIS through the TRL stages. Risk assessment involves identifying potential technical, operational, and financial risks associated with development. Mitigation strategies may include developing contingency plans, conducting risk assessments, and implementing risk reduction measures throughout the development process.

3. Practical Implementation and Evaluation

3.1. Prototyping and Testing

3.1.1. Prototype Development

Developing prototypes is a key step in advancing BIS through the TRL stages. Prototypes should be designed to test specific components and functionalities of BIS. This includes building small-scale models and conducting laboratory tests to validate design principles. Each prototype iteration should incorporate lessons learned from previous tests and address any identified issues.

3.1.2. Testing and Validation

Testing BIS prototypes in relevant environments is crucial for assessing performance and reliability. This includes evaluating prototypes in simulated operational conditions and conducting field tests to assess real-world performance. Testing should be rigorous and comprehensive, covering various scenarios to ensure that BIS meets all operational requirements.

3.2. Integration and System Demonstration

3.2.1. System Integration

Integrating BIS components into a cohesive system is essential for demonstrating functionality and performance. This involves assembling and configuring components, ensuring compatibility, and optimizing system performance. Integration should be followed by system-level testing to validate that all components work together effectively.

3.2.2. System Demonstration

Demonstrating the BIS system in operational environments is a critical step towards achieving higher TRLs. System demonstrations should showcase the technology's capabilities and performance in real-world scenarios. Successful demonstrations provide evidence of BIS's readiness for deployment and operational use.

4. Expansive Perspective on TRL Advancement

4.1. Long-Term Implications for BIS Technology

4.1.1. Impact on Industry and Defense

Advancing BIS to higher TRLs has significant implications for the industry and defense sectors. A fully developed BIS system could enhance personal protection, revolutionize security measures, and provide new capabilities for law enforcement and military applications. Achieving higher TRLs also opens opportunities for commercialization and integration into various security solutions.

4.1.2. Future Research and Development
Reaching higher TRLs for BIS sets the stage for future research and development. Continuous innovation and improvement are necessary to address emerging challenges, incorporate new technologies, and enhance performance. Future R&D efforts will focus on refining BIS technology, exploring new applications, and maintaining technological leadership.

4.2. Strategic Planning and Collaboration

4.2.1. Collaborative Efforts
Collaboration with research institutions, industry partners, and government agencies is essential for advancing BIS through the TRL stages. Strategic partnerships can provide additional resources, expertise, and funding. Collaborative efforts also facilitate knowledge sharing and accelerate technology development.

4.2.2. Strategic Roadmap Development
Developing a strategic roadmap is crucial for guiding BIS through the TRL stages. The roadmap should outline clear objectives, milestones, and timelines for each TRL level. Regular reviews and updates to the roadmap ensure that development remains on track and adapts to changing requirements and opportunities.

Conclusion

Technological Readiness Levels (TRLs) provide a structured approach for assessing and advancing the Bullet Inversion System (BIS) through its development stages. By understanding and applying TRL metrics, stakeholders can effectively evaluate BIS's current status, plan development stages, and work towards achieving higher TRLs. This systematic approach ensures that BIS technology progresses from concept to fully operational deployment, addressing technical, operational, and financial challenges along the way. The application of TRLs not only enhances the development process but also paves the way for successful integration and commercialization of advanced protective technologies.

Cross-Disciplinary Collaboration: Bridging Physics, Engineering, and Materials Science

1. The Importance of Cross-Disciplinary Collaboration

1.1. Integrating Diverse Expertise

Cross-disciplinary collaboration is crucial for advancing complex technologies and addressing multifaceted challenges. In fields such as the Bullet Inversion System (BIS), which involves intricate interplay between electromagnetic fields, materials science, and engineering principles, collaboration across disciplines brings together diverse expertise that can lead to more innovative solutions. Each discipline contributes unique insights that collectively enhance the overall understanding and development of the technology.

- **Physics**: Physics provides the foundational principles governing electromagnetic fields, plasma dynamics, and material interactions. Theoretical models, such as those describing Lorentz force laws and Maxwell's equations, are essential for predicting and manipulating physical phenomena in BIS technology.

- **Engineering**: Engineering translates theoretical principles into practical applications. It involves designing and optimizing systems that can implement the physical concepts developed by physicists. Engineering expertise is crucial in developing prototypes, managing power requirements, and ensuring system reliability and efficiency.

- **Materials Science**: Materials science focuses on the properties and behavior of materials under different conditions. For BIS, materials scientists are essential for selecting and developing materials that can withstand high-energy environments, ensuring durability and performance of components such as magnetic field generators and plasma systems.

1.2. Driving Innovation Through Integration

Collaboration across these disciplines fosters innovation by combining theoretical knowledge with practical application. For instance, the integration of advanced superconducting materials in BIS systems requires both a deep understanding of materials science and the ability to engineer effective cooling and power management systems. Physics provides the theoretical framework for how superconductors interact with magnetic fields, while engineering applies this knowledge to create functional and reliable devices.

- **Complex Problem Solving**: Complex technologies like BIS require solving problems that span multiple domains. Cross-disciplinary teams can address these challenges more effectively by leveraging each discipline's strengths. For example, understanding the interplay between electromagnetic fields and plasma requires both theoretical and experimental insights that span physics and engineering.
- **Innovative Solutions**: Collaboration often leads to novel approaches and solutions that may not emerge within the confines of a single discipline. For example, combining insights from materials science and engineering can lead to the development of new materials with enhanced properties, which in turn can improve the performance of BIS technology.

2. Fostering Interdisciplinary Research Teams

2.1. Building Effective Teams

Creating effective interdisciplinary research teams involves careful planning and coordination. Successful teams integrate members from different disciplines, each bringing their specialized knowledge to the table. Key factors in building these teams include:

- **Clear Objectives**: Defining clear research objectives helps align the diverse expertise of team members towards a common goal. For BIS, objectives might include developing new materials for magnetic field generation or designing innovative cooling systems for superconductors.

- **Complementary Skills**: Selecting team members with complementary skills ensures that all aspects of the technology are addressed. For example, a team working on BIS might include experts in electromagnetic theory, plasma physics, materials science, and systems engineering.

- **Communication and Collaboration**: Effective communication is essential for interdisciplinary collaboration. Team members must be able to share their insights and knowledge clearly and work together to integrate their findings. Regular meetings and collaborative tools can facilitate communication and ensure that all perspectives are considered.

2.2. Promoting Interdisciplinary Education

Education plays a critical role in fostering interdisciplinary research. Programs that emphasize cross-disciplinary training prepare future researchers to work effectively in collaborative environments. Strategies for promoting interdisciplinary education include:

- **Integrated Curricula**: Developing curricula that combine elements from multiple disciplines helps students gain a broad understanding of complex problems. For example, a program that integrates physics, engineering, and materials science can prepare students to tackle advanced technologies like BIS.

- **Collaborative Projects**: Encouraging students to work on collaborative projects provides practical experience in interdisciplinary research. Projects that require input from multiple disciplines help students understand the value of different perspectives and develop the skills needed for effective collaboration.

- **Interdisciplinary Workshops and Seminars**: Hosting workshops and seminars that bring together experts from various fields can stimulate interest in interdisciplinary research and facilitate knowledge exchange. These events provide opportunities for researchers to learn about each other's work and identify potential areas for collaboration.

2.3. Facilitating Research Partnerships

Establishing partnerships between academic institutions, industry, and government agencies can enhance interdisciplinary research efforts. Partnerships provide access to resources, expertise, and funding, which are crucial for advancing complex technologies. Strategies for fostering research partnerships include:

- **Industry-Academia Collaborations**: Collaborations between academic institutions and industry can accelerate the development of new technologies. Industry partners provide practical insights and resources, while academic researchers contribute theoretical knowledge and advanced research capabilities.

- **Government Funding and Support**: Government agencies often fund research initiatives that address national priorities or advance key technologies. Securing government support for interdisciplinary research can provide essential funding and resources for developing technologies like BIS.

- **Research Consortia**: Forming research consortia with multiple stakeholders allows for pooling of expertise and resources. Consortia enable large-scale projects that require diverse knowledge and capabilities, fostering collaboration across institutions and disciplines.

3. Practical Applications and Challenges

3.1. Addressing Practical Challenges

Interdisciplinary collaboration presents practical challenges that must be addressed to ensure effective teamwork:

- **Differing Terminologies**: Different disciplines often use specialized terminologies that can create barriers to communication. Establishing a common language or glossary can help bridge these gaps and facilitate understanding.

- **Cultural Differences**: Disciplines may have different research cultures and approaches. Fostering a collaborative culture that values diverse perspectives and approaches is essential for effective teamwork.

- **Resource Allocation**: Coordinating resources across disciplines can be challenging. Clear planning and management strategies are needed to ensure that resources are allocated effectively and that all team members have access to the tools and support they need.

3.2. Leveraging Collaborative Success

Successful interdisciplinary collaboration can lead to significant advancements and breakthroughs:

- **Enhanced Innovation**: Combining expertise from multiple disciplines often leads to innovative solutions that address complex problems more effectively. For BIS, interdisciplinary collaboration can drive advancements in materials, systems design, and performance.

- **Broader Impact**: Technologies developed through interdisciplinary research often have broader applications and impact. For example, advancements in BIS technology may lead to improvements in other areas, such as energy storage or materials science.

4. Conclusion

Cross-disciplinary collaboration is essential for advancing complex technologies like the Bullet Inversion System (BIS). By integrating expertise from physics, engineering, and materials science, researchers can address multifaceted challenges and drive innovation. Building effective interdisciplinary research teams, promoting cross-disciplinary education, and fostering research partnerships are key strategies for successful collaboration. Addressing practical challenges and leveraging collaborative success can lead to significant advancements and broader impact, ultimately advancing technology and addressing critical needs in various fields.

Long-Term Vision and Goals: Shaping the Future of BIS Technology

1. Setting Long-Term Research and Development Goals

1.1. Defining the Scope and Objectives

Establishing long-term research and development (R&D) goals for the Bullet Inversion System (BIS) involves articulating a clear vision of the technology's future, identifying key milestones, and setting strategic objectives that guide the evolution of the system. The scope of these goals encompasses various dimensions, including technological advancements, practical applications, and societal impact.

- **Technological Advancements**: One of the primary goals is to advance the fundamental technology behind BIS. This includes improving the efficiency and effectiveness of magnetic field generators, refining plasma generation techniques, and enhancing the overall integration of these components. Achieving these advancements involves incremental progress in each technological area, culminating in significant breakthroughs that push the boundaries of what BIS can accomplish.

- **Practical Applications**: Setting long-term goals also requires envisioning the practical applications of BIS technology. This involves identifying potential use cases, ranging from personal protection to military and law enforcement applications. By defining specific scenarios where BIS can be effectively deployed, researchers can focus on developing solutions tailored to these needs, ensuring that the technology delivers real-world benefits.

- **Societal Impact**: Long-term goals should also consider the broader societal impact of BIS technology. This includes addressing ethical, legal, and environmental concerns, and ensuring that the technology contributes positively to society. Goals in this area might involve developing strategies for responsible deployment, mitigating potential risks, and maximizing the positive impact of BIS on public safety and security.

1.2. Strategic Planning and Milestones

To achieve long-term goals, it is essential to develop a strategic plan with clearly defined milestones. These milestones serve as checkpoints that guide the progress of R&D efforts and ensure that the technology evolves in line with the overarching vision.

- **Short-Term Milestones**: Short-term milestones might include the development of functional prototypes, successful laboratory testing, and preliminary field trials. These milestones provide early indicators of progress and help identify potential challenges that need to be addressed.

- **Intermediate Milestones**: Intermediate milestones could involve scaling up prototypes, conducting extensive field testing, and refining the technology based on feedback. These milestones represent significant steps toward achieving the long-term vision and require ongoing evaluation and adjustment of strategies.

- **Long-Term Milestones**: Long-term milestones include the commercialization of BIS technology, widespread adoption in various sectors, and achieving significant technological advancements. These milestones represent the culmination of R&D efforts and the realization of the technology's full potential.

2. Vision for Future Iterations and Improvements

2.1. Continuous Improvement and Innovation

The vision for future iterations of BIS technology involves a commitment to continuous improvement and innovation. This approach ensures that the technology remains at the forefront of advancements and continues to meet evolving needs and challenges.

- **Enhancing Performance**: Future iterations of BIS should focus on enhancing performance metrics, such as the efficiency of magnetic field generation, the stability of plasma fields, and the overall effectiveness of the system in real-world scenarios. This involves ongoing research to identify and implement improvements that boost performance and address any limitations.

- **Incorporating New Technologies**: Integrating emerging technologies into BIS can drive significant improvements. For example, advancements in materials science, energy storage, and computational modeling can enhance the capabilities of BIS systems. By staying abreast of technological trends and incorporating relevant innovations, future iterations can achieve higher levels of performance and functionality.

- **User-Centric Design**: Future iterations should also prioritize user-centric design, focusing on making BIS technology more accessible and user-friendly. This involves incorporating feedback from end-users, simplifying operation, and improving ergonomics to ensure that the technology meets the needs of its intended users.

2.2. Expanding Applications and Use Cases

As BIS technology evolves, expanding its applications and use cases is a key aspect of the long-term vision. This involves exploring new domains where BIS can provide value and addressing the unique requirements of these applications.

- **Military and Law Enforcement**: Continued development of BIS for military and law enforcement applications involves adapting the technology to meet the specific demands of these fields. This might include enhancing the system's capability to handle different types of threats, improving integration with existing protective gear, and addressing operational challenges in diverse environments.

- **Civilian Applications**: Exploring civilian applications of BIS can open up new opportunities for the technology. Potential areas include personal safety, crowd control, and emergency response. By identifying and developing solutions for these applications, BIS technology can contribute to broader societal benefits.

- **International Markets**: Expanding BIS technology to international markets involves understanding and addressing regional needs and regulatory requirements. This might include adapting the technology to meet different standards and preferences, as well as developing strategies for global distribution and support.

2.3. Addressing Future Challenges and Opportunities

The long-term vision for BIS technology also involves anticipating and addressing future challenges and opportunities. This proactive approach ensures that the technology remains resilient and adaptable to changing circumstances.

- **Technological Challenges**: Future challenges may include evolving threats, advances in counter-technologies, and changes in materials and energy requirements. Addressing these challenges requires ongoing research and development to stay ahead of emerging trends and ensure that BIS technology remains effective and relevant.

- **Regulatory and Ethical Considerations**: Navigating regulatory and ethical considerations is crucial for the future of BIS technology. This involves ensuring compliance with regulations, addressing privacy and security concerns, and engaging with stakeholders to build trust and support.

- **Sustainability and Environmental Impact**: Future iterations of BIS should also focus on sustainability and minimizing environmental impact. This involves developing energy-efficient solutions, using environmentally friendly materials, and implementing recycling and waste management strategies.

3. Conclusion

Setting long-term research and development goals for the Bullet Inversion System (BIS) involves defining a clear vision, establishing strategic milestones, and committing to continuous improvement and innovation. By focusing on technological advancements, practical applications, and societal impact, researchers can guide the evolution of BIS technology toward achieving its full potential. Future iterations should prioritize performance enhancement, incorporation of new technologies, and user-centric design while expanding applications and addressing emerging challenges. This proactive approach ensures that BIS technology remains at the forefront of advancements and continues to deliver value in diverse contexts.

Regulatory Compliance for Bullet Inversion System (BIS)

1. Navigating Regulatory Requirements

1.1. Understanding the Regulatory Landscape

Navigating the regulatory landscape for the Bullet Inversion System (BIS) involves a comprehensive understanding of various regulations, standards, and guidelines that govern the development, deployment, and operation of such advanced technologies. The BIS intersects with multiple regulatory domains, including safety, performance, environmental impact, and more. Therefore, it's crucial to identify and interpret relevant regulations to ensure compliance.

- **National Regulations**: In many countries, the development and deployment of advanced protective technologies like BIS are subject to national regulations. These regulations may include standards for safety, electromagnetic emissions, and materials used. For instance, in the United States, agencies such as the Federal Aviation Administration (FAA) and the National Institute of Standards and Technology (NIST) may have relevant guidelines, depending on the BIS's applications. Similarly, the European Union has stringent regulations under the CE marking process that BIS would need to comply with.

- **International Standards**: Compliance with international standards is essential, especially if BIS technology is intended for global markets. Organizations like the International Electrotechnical Commission (IEC) and the International Organization for Standardization (ISO) provide guidelines and standards for various aspects of technology, including electromagnetic compatibility (EMC) and safety performance. Adhering to these standards ensures that BIS meets global expectations and facilitates international trade.

- **Sector-Specific Regulations**: Different sectors may have specific regulatory requirements. For example, if BIS is used in military or law enforcement applications, it must comply with sector-specific standards and regulations, such as those set by the Department of Defense (DoD) in the U.S. or similar bodies in other countries.

Understanding and addressing these sector-specific regulations is crucial for successful deployment.

1.2. Regulatory Approval Processes

The process of obtaining regulatory approval for BIS technology typically involves several stages, including pre-market assessment, testing, and certification. Each stage requires thorough documentation and adherence to regulatory requirements.

- **Pre-Market Assessment**: Before BIS can be introduced to the market, it must undergo a pre-market assessment to ensure that it meets all regulatory requirements. This may involve submitting detailed technical documentation, including design specifications, safety analyses, and performance data. Regulatory bodies review this information to determine if the technology meets the required standards.

- **Testing and Certification**: BIS technology must be tested to ensure compliance with safety and performance standards. This may involve independent testing by accredited laboratories or certification bodies. For example, electromagnetic compatibility (EMC) testing ensures that the BIS does not interfere with other electronic devices and that it operates as intended. Certification indicates that the technology meets the necessary standards and can be marketed and used legally.

- **Ongoing Compliance**: Regulatory compliance is not a one-time process but an ongoing responsibility. Once BIS is in the market, it must continue to meet regulatory requirements throughout its lifecycle. This includes regular inspections, updating compliance documentation, and addressing any changes in regulations or standards.

2. Ensuring Compliance with Safety and Performance Standards

2.1. Safety Standards

Ensuring compliance with safety standards is critical for the BIS to protect users and the environment. Various safety standards address different aspects of technology, from electromagnetic emissions to material safety.

- **Electromagnetic Emissions**: BIS technology must comply with standards related to electromagnetic emissions to prevent interference with other electronic devices. Regulations like the Federal Communications Commission (FCC) Part 15 in the U.S. set limits on electromagnetic interference (EMI). Ensuring that BIS adheres to these limits is essential for both safety and regulatory compliance.

- **Material Safety**: The materials used in BIS must meet safety standards to prevent harm to users. This includes assessing the toxicity of materials, ensuring that they are non-reactive, and complying with regulations regarding material handling and disposal. For example, regulations may require that materials used in BIS do not release harmful substances during use or in the event of a malfunction.

- **Operational Safety**: BIS technology must be designed to operate safely under various conditions. This involves incorporating safety features such as automatic shutdown mechanisms, fail-safes, and user warnings. Compliance with operational safety standards ensures that BIS functions reliably and does not pose a risk to users.

2.2. Performance Standards

Meeting performance standards is equally important for BIS to ensure that it functions as intended and delivers the expected benefits. Performance standards cover various aspects, including efficiency, reliability, and durability.

- **Efficiency**: BIS must meet efficiency standards to ensure optimal performance. This involves evaluating the system's energy consumption, effectiveness in generating magnetic fields, and overall operational efficiency. Compliance with efficiency standards helps optimize performance while minimizing energy use and operational costs.

- **Reliability**: Reliability standards ensure that BIS performs consistently over time. This includes assessing the technology's ability to withstand operational stresses, maintain performance under various conditions, and provide reliable protection. Reliability testing may involve stress tests, durability tests, and long-term performance evaluations.

- **Durability**: BIS must be durable to ensure long-term effectiveness and safety. Durability standards address factors such as resistance to wear and tear, environmental conditions, and maintenance requirements. Ensuring compliance with durability standards helps guarantee that BIS remains functional and effective throughout its intended lifespan.

2.3. Documentation and Reporting

Accurate documentation and reporting are essential for regulatory compliance. Detailed records of design, testing, certification, and ongoing compliance activities must be maintained and readily accessible.

- **Design Documentation**: Comprehensive design documentation includes technical specifications, safety analyses, and performance data. This documentation provides a clear record of the design process and demonstrates compliance with regulatory requirements.

- **Testing Reports**: Testing reports document the results of performance and safety tests. These reports are crucial for demonstrating compliance with regulatory standards and addressing any issues identified during testing.

- **Compliance Records**: Ongoing compliance records include updates to regulations, inspection reports, and any corrective actions taken. Maintaining these records ensures that BIS continues to meet regulatory requirements and facilitates any necessary updates or adjustments.

3. Conclusion

Navigating regulatory requirements for the Bullet Inversion System (BIS) involves a comprehensive understanding of national, international, and sector-specific regulations. Ensuring compliance with safety and performance standards is essential for the successful development and deployment of BIS technology. By addressing regulatory approval processes, adhering to safety and performance standards, and maintaining accurate documentation, developers can ensure that BIS meets all regulatory requirements and delivers its intended benefits. Ongoing compliance efforts and proactive engagement with regulatory bodies help maintain the technology's effectiveness and safety, paving the way for successful market integration and deployment.

User Interface and Experience Design for Bullet Inversion System (BIS)

1. Designing an Intuitive User Interface for BIS Operation

1.1. Principles of Intuitive Design

Designing an intuitive user interface (UI) for the Bullet Inversion System (BIS) involves creating an interface that is easy to understand, navigate, and use effectively. The goal is to minimize the learning curve for users and ensure that the BIS operates seamlessly in high-pressure situations. The principles of intuitive design include simplicity, consistency, and responsiveness.

- **Simplicity**: An intuitive UI should present only the essential information and controls needed for operation. This involves avoiding clutter and focusing on the primary functions of the BIS. Simple, clear icons, labels, and instructions help users quickly grasp the system's functionality. For example, a well-designed dashboard for BIS might include a straightforward control panel with large, easily identifiable buttons for activating the system, adjusting settings, and monitoring status.

- **Consistency**: Consistency in design ensures that users can predict how different elements of the UI will behave based on their previous interactions. This includes using a uniform layout, color scheme, and terminology throughout the interface. For instance, if the BIS interface uses specific icons for power and status indicators, these icons should be consistently applied across all screens and functions.

- **Responsiveness**: The UI must be responsive to user inputs, providing immediate feedback to confirm that actions have been received and processed. This includes visual feedback such as button highlights or status updates, as well as auditory cues if appropriate. For instance, when a user activates a function on the BIS interface, a confirmation sound or visual indicator ensures that the command has been executed.

1.2. Interaction Design

Interaction design focuses on how users interact with the BIS interface and how these interactions are structured to enhance usability. Key aspects of interaction design include input methods, control mechanisms, and feedback.

- **Input Methods**: The BIS interface should support various input methods to accommodate different user preferences and scenarios. This may include touch screens, physical buttons, and voice commands. Designing for multiple input methods ensures that the BIS can be operated efficiently in diverse environments and by users with varying levels of technical expertise.

- **Control Mechanisms**: Controls should be logically organized and easy to access. Critical functions should be prominently placed, while less frequently used features can be nested within menus or sub-systems. For example, frequently used controls for activating or deactivating the BIS might be placed on the main screen, while advanced settings and diagnostics could be accessed through a secondary menu.

- **Feedback Mechanisms**: Providing clear feedback to users is crucial for effective operation. This includes visual indicators such as status lights or progress bars, as well as textual messages and alerts. For instance, if a user adjusts a setting on the BIS, a real-time display of the setting's current value and a confirmation message can help ensure that the adjustment is correct.

2. Enhancing User Experience and Accessibility

2.1. User Experience (UX) Enhancements

Enhancing user experience involves creating a UI that not only meets functional requirements but also provides a positive and engaging experience for users. Key considerations for enhancing UX include personalization, usability testing, and user support.

- **Personalization**: Allowing users to personalize the BIS interface can improve their experience by tailoring the system to their preferences. This might include customizable dashboards, adjustable control layouts, and personalized alerts. For example, users could choose their preferred display mode (e.g., light or dark theme) or rearrange controls to suit their workflow.

- **Usability Testing**: Conducting usability testing with real users is essential for identifying and addressing potential issues in the BIS interface. This involves observing users as they interact with the system, gathering feedback, and analyzing their behavior to identify areas for improvement. Usability testing helps ensure that the BIS interface is intuitive, efficient, and effective.

- **User Support**: Providing comprehensive user support, including tutorials, help screens, and troubleshooting guides, can enhance the overall user experience. This support should be easily accessible from within the BIS interface, allowing users to quickly find assistance when needed. For instance, an integrated help feature with searchable FAQs and step-by-step guides can assist users in resolving issues or learning new features.

2.2. Accessibility Considerations

Ensuring accessibility is crucial for making the BIS interface usable by individuals with diverse abilities and needs. Key accessibility considerations include accommodating visual, auditory, and motor impairments.

- **Visual Accessibility**: The BIS interface should be designed with visual accessibility in mind, including high-contrast text, adjustable font sizes, and color-blind friendly palettes. Features such as screen readers and magnification tools should also be supported to assist users with visual impairments. For instance, incorporating text-to-speech functionality can help users with low vision navigate the interface.

- **Auditory Accessibility**: For users with hearing impairments, the BIS interface should provide alternative means of communication. This includes visual alerts and notifications in place of auditory cues. For example, if the BIS uses sound alerts for status updates, visual indicators such as flashing lights or on-screen messages can be provided as alternatives.

- **Motor Accessibility**: The interface should accommodate users with motor impairments by offering alternative input methods and adjustable control sensitivity. This might include support for adaptive input devices, customizable control layouts, and options for adjusting button size and spacing. Ensuring that controls are accessible and easy to operate is key to accommodating users with varying motor abilities.

3. Integration with Other Technologies

Integrating the BIS interface with other technologies can enhance its functionality and improve the overall user experience. This includes integration with mobile devices, cloud services, and data analytics.

- **Mobile Integration**: Allowing users to interact with the BIS through mobile devices can provide greater flexibility and convenience. This might involve developing mobile apps or web-based interfaces that enable remote monitoring and control of the BIS. Mobile integration can be particularly useful for users who need to access the BIS while on the move or in different locations.

- **Cloud Services**: Integrating BIS with cloud services can provide additional features such as data storage, analytics, and remote updates. Cloud-based solutions can facilitate real-time data synchronization, allowing users to access and analyze BIS performance data from anywhere. This integration also enables the deployment of software updates and enhancements without requiring direct access to the BIS hardware.

144

- **Data Analytics**: Leveraging data analytics can help users gain insights into the performance and operation of the BIS. This includes analyzing usage patterns, identifying trends, and detecting potential issues. Data-driven insights can inform design improvements, optimize system performance, and enhance overall user experience.

4. Conclusion

Designing an intuitive user interface and enhancing user experience for the Bullet Inversion System (BIS) requires a thoughtful approach to interaction design, accessibility, and integration with other technologies. By focusing on simplicity, consistency, and responsiveness in the UI design, and by addressing accessibility considerations and providing comprehensive user support, developers can create a system that is both effective and user-friendly. Integrating the BIS with mobile devices, cloud services, and data analytics further enhances its functionality and provides users with a seamless and engaging experience. Ultimately, a well-designed UI and UX will contribute to the successful deployment and operation of the BIS, ensuring that it meets the needs of its users and delivers its intended benefits.

Robustness and Reliability Testing for Bullet Inversion System (BIS)

1. Ensuring BIS Reliability Under Diverse Conditions

1.1. Defining Reliability in BIS

Reliability in the context of the Bullet Inversion System (BIS) refers to the system's ability to perform its intended function consistently and effectively over time, under various operational conditions. Reliability encompasses several factors, including operational dependability, consistency in performance, and resistance to failure. Ensuring BIS reliability involves rigorous testing and validation across different scenarios to verify that the system performs as expected under diverse conditions.

1.2. Operational Scenarios for Reliability Testing

To ensure BIS reliability, it is crucial to test the system under a range of operational scenarios that mimic real-world conditions. These scenarios include:

- **Environmental Extremes**: BIS should be tested in extreme environmental conditions, such as high temperatures, low temperatures, humidity, and dust. This testing ensures that the system can operate reliably in varied climates and environmental settings. For instance, high-temperature tests simulate conditions in hot climates, while low-temperature tests assess performance in cold environments.

- **Operational Stress**: Testing under stress conditions involves subjecting the BIS to intense operational demands, such as continuous use, rapid activation and deactivation, and extended operational periods. This helps identify potential issues related to system performance under sustained or extreme use.

- **Interference and Electromagnetic Conditions**: Given that BIS relies on electromagnetic principles, testing for electromagnetic interference (EMI) and compatibility is essential. The system should be evaluated for its ability to function reliably in the presence of various sources of EMI, such as electronic devices, communication systems, and other electromagnetic fields.

2. Testing for Robustness and Durability

2.1. Assessing Robustness

Robustness testing evaluates how well the BIS withstands external forces and impacts that could potentially compromise its functionality. Key aspects of robustness testing include:

- **Mechanical Stress Testing**: The BIS should be subjected to mechanical stress tests, including impacts, vibrations, and shock tests. These tests simulate scenarios where the system may experience physical stress, such as being dropped, bumped, or subjected to vibrations. Robustness testing helps identify vulnerabilities in the system's physical construction and ensures that it can endure physical abuse without failure.

- **Structural Integrity**: Ensuring the structural integrity of the BIS involves evaluating the durability of its housing, components, and connections. This includes tests for resistance to cracks, deformation, and component dislodgment. The BIS's casing, connectors, and internal components must be designed to withstand rough handling and operational stress.

- **Ingress Protection**: Testing for ingress protection assesses the BIS's resistance to environmental elements such as water, dust, and debris. This involves exposing the system to water spray, immersion, and dust particles to verify that it maintains functionality and does not suffer from ingress-related damage.

2.2. Evaluating Durability

Durability testing focuses on the long-term performance and longevity of the BIS. It involves assessing how the system holds up over extended periods of use and exposure to various conditions. Key aspects of durability testing include:

- **Wear and Tear Testing**: This testing evaluates how the BIS components endure over time with regular use. Components are subjected to simulated wear and tear to identify potential points of failure. For instance, repeated activation cycles, mechanical movements, and component interactions are tested to assess their impact on system durability.

- **Component Lifespan**: Evaluating the lifespan of individual components, such as electronic circuits, magnetic elements, and cooling systems, is crucial. This involves accelerated aging tests to simulate long-term usage and determine when components are likely to degrade or fail.

- **Maintenance and Repair**: Durability testing also considers the ease of maintenance and repair. The BIS should be designed for straightforward servicing, with components that can be replaced or repaired without extensive disassembly. This aspect ensures that the system remains functional and cost-effective over its operational life.

3. Implementing Testing Protocols

3.1. Test Planning and Execution

Developing a comprehensive testing plan is essential for effective robustness and reliability testing. The plan should include:

- **Test Objectives**: Clearly define the objectives of each test, including what aspects of robustness or reliability are being evaluated. Objectives should align with the specific requirements and intended use of the BIS.

- **Test Procedures**: Outline detailed procedures for conducting each test, including equipment, methods, and conditions. Consistency in test procedures ensures reliable and comparable results.

- **Data Collection and Analysis**: Implement systematic data collection methods to record test results. This includes quantitative measurements, qualitative observations, and any incidents of failure. Analyzing this data helps identify trends, potential weaknesses, and areas for improvement.

3.2. Iterative Testing and Improvement

Robustness and reliability testing is an iterative process. After initial testing, any identified issues should be addressed through design improvements and retesting. This iterative approach ensures that the BIS evolves to meet robustness and reliability standards.

- **Feedback Integration**: Incorporate feedback from testing into the design process. For example, if a specific component fails under stress testing, redesigning that component to enhance its durability can lead to improved overall system performance.
- **Continuous Monitoring**: Even after initial testing and deployment, continuous monitoring of the BIS in real-world conditions provides valuable insights into its long-term reliability. Gathering performance data from operational use can guide future updates and refinements.

4. Conclusion

Robustness and reliability testing are critical for ensuring that the Bullet Inversion System (BIS) performs consistently and effectively under diverse conditions. By defining reliability, testing under various operational scenarios, and assessing robustness and durability, developers can identify potential issues and enhance system performance. Implementing comprehensive testing protocols and adopting an iterative approach to design and improvement are essential for achieving a robust and reliable BIS. Through rigorous testing and continuous monitoring, the BIS can be optimized to meet the highest standards of performance and durability, ensuring its effectiveness in real-world applications.

Cultural and Geopolitical Considerations for Bullet Inversion System (BIS)

1. Understanding Cultural Attitudes Towards Defense Technology

1.1. Cultural Attitudes and Perceptions

Cultural attitudes towards defense technology, including advanced systems like the Bullet Inversion System (BIS), vary significantly across different societies and regions. Understanding these cultural attitudes is crucial for the successful deployment and acceptance of BIS technology.

- **Security and Safety Priorities**: In some cultures, personal and national security is a paramount concern, leading to a more favorable view of advanced defense technologies. For example, in countries facing high security threats or experiencing ongoing conflicts, there may be strong public support for innovative defense solutions that enhance protection and safety. Conversely, in societies with a lower perceived threat level, the emphasis may be on reducing military expenditure and focusing on social and economic development, which might lead to skepticism towards advanced defense technologies.

- **Ethical and Moral Considerations**: Different cultures have varying ethical perspectives on the use of defense technology. In some societies, there is a strong emphasis on the moral implications of using high-tech defense systems, including concerns about privacy, the potential for misuse, and the impact on civilian life. Engaging with these cultural perspectives requires addressing ethical concerns transparently and incorporating public input into the development and deployment of BIS technology.

- **Historical Context and Traditions**: Historical experiences and traditions influence how new technologies are perceived. Cultures with a history of technological advancement may be more accepting of new defense technologies, while others may approach them with caution due to past conflicts or negative experiences with technology. Understanding these historical contexts helps in framing the BIS technology in a way that aligns with cultural values and historical narratives.

1.2. Engaging with Diverse Cultural Perspectives

- **Community Involvement**: Engaging with local communities and stakeholders is essential for understanding and addressing cultural attitudes. This can be achieved through public consultations, focus groups, and community outreach programs. Listening to and addressing concerns from various cultural and social groups helps in tailoring the BIS technology to meet cultural expectations and gain broader acceptance.

- **Education and Awareness**: Developing educational programs and awareness campaigns that explain the benefits and limitations of BIS technology can help shift cultural attitudes. Providing clear, accessible information about how BIS works, its safety features, and its potential benefits can foster a more informed and balanced view among different cultural groups.

- **Tailoring Communication**: Communication strategies should be adapted to suit cultural norms and values. For example, in cultures where traditional media is predominant, leveraging television, radio, and print media might be more effective, while in technology-savvy cultures, digital platforms and social media might play a larger role.

2. Geopolitical Implications of Widespread BIS Use

2.1. Strategic Balance and Arms Race

The introduction and widespread adoption of BIS technology could have significant implications for global strategic balance and arms races.

- **Shifts in Power Dynamics**: The deployment of advanced defense systems like BIS could alter power dynamics among nations. Countries with access to such technology might gain a strategic advantage, potentially leading to shifts in global power balances. This could provoke reactions from other nations, including efforts to develop or acquire similar technologies, thereby influencing the geopolitical landscape.
- **Arms Race Dynamics**: The availability of advanced defense systems might trigger an arms race, where countries feel compelled to develop or acquire counter-technologies. This dynamic can lead to increased military spending and heightened tensions between nations, as countries seek to maintain or enhance their security and strategic positions.

2.2. International Relations and Cooperation

The global distribution of BIS technology could influence international relations and cooperation in various ways.

- **Alliances and Partnerships**: Nations with advanced BIS technology might form alliances or partnerships based on mutual interests in security and defense. These alliances could lead to collaborative efforts in technology development, joint defense initiatives, and shared intelligence. Conversely, nations without access to BIS technology might seek alternative partnerships or support from other allies to balance the perceived technological gap.

- **Arms Control Agreements**: The proliferation of advanced defense technologies could prompt discussions and negotiations regarding arms control agreements. International bodies and treaties might be established or updated to address the implications of BIS technology, aiming to prevent an uncontrolled arms race and promote stability and transparency in defense technology.

- **Global Security Frameworks**: The integration of BIS technology into national defense strategies could influence global security frameworks. International organizations and security alliances might need to adapt their strategies and policies to accommodate new technological realities, ensuring that BIS technology is managed responsibly and does not undermine global security efforts.

2.3. Impact on Global Trade and Economics

The widespread adoption of BIS technology could have economic implications, affecting global trade and industry.

- **Economic Opportunities**: Countries that lead in BIS technology development might benefit economically through exports, technology transfers, and defense contracts. This can create economic opportunities and stimulate growth in technology sectors, contributing to national economic development.
- **Trade Barriers and Sanctions**: Conversely, the proliferation of advanced defense technologies might lead to trade barriers or sanctions. Nations might impose restrictions on the export or transfer of BIS technology to prevent proliferation or maintain strategic advantages. This could impact global trade dynamics and influence the availability and cost of BIS technology.

3. Addressing Geopolitical and Cultural Challenges

3.1. Strategic Diplomacy

Engaging in strategic diplomacy is essential for managing the geopolitical implications of BIS technology. This involves:

- **Dialogue and Negotiation**: Establishing dialogue with other nations to address concerns and negotiate terms for technology sharing and arms control. Transparent communication and collaborative efforts can help mitigate tensions and promote mutual understanding.
- **International Collaboration**: Participating in international forums and collaborative projects to address global security challenges and ensure that BIS technology is used responsibly. This includes contributing to international standards and agreements that govern the development and deployment of advanced defense systems.

3.2. Cultural Sensitivity and Adaptation

Adapting BIS technology and deployment strategies to cultural contexts is crucial for gaining acceptance and avoiding potential conflicts.

- **Local Customization**: Customizing BIS technology to fit cultural norms and preferences can enhance acceptance. This might involve modifying features, user interfaces, or communication strategies to align with cultural expectations.
- **Ethical Considerations**: Addressing ethical concerns and ensuring that BIS technology aligns with international human rights standards. This includes conducting impact assessments and engaging with ethicists and cultural experts to ensure responsible deployment.

4. Conclusion

Cultural and geopolitical considerations play a critical role in the development, deployment, and acceptance of advanced defense technologies like the Bullet Inversion System (BIS). Understanding cultural attitudes towards defense technology and addressing geopolitical implications is essential for successful implementation. Engaging with diverse cultural perspectives, managing strategic and economic impacts, and fostering international cooperation and diplomatic efforts are key to navigating these complex challenges. By addressing these considerations thoughtfully and proactively, stakeholders can enhance the acceptance and effectiveness of BIS technology while contributing to global security and stability.

Scalability and Mass Production of Bullet Inversion Systems (BIS)

1. Planning for Scalable Manufacturing Processes

1.1. Principles of Scalability

Scalability in manufacturing refers to the ability to increase production volumes efficiently without compromising quality or increasing costs disproportionately. For advanced technologies like the Bullet Inversion System (BIS), achieving scalability involves careful planning and strategic design choices.

- **Modular Design**: One approach to scalability is adopting a modular design for BIS components. By creating standardized modules that can be assembled or disassembled, manufacturers can easily adjust production volumes. Modular designs facilitate flexibility in production lines, allowing for rapid adjustments to accommodate changing demand or technological updates. This approach also simplifies maintenance and upgrades, as individual modules can be replaced or improved without overhauling the entire system.

- **Automation and Robotics**: Integrating automation and robotics into the manufacturing process can significantly enhance scalability. Automated systems, including robotic arms and conveyor belts, streamline repetitive tasks, reduce human error, and increase production speed. Advanced automation solutions can be programmed to handle different BIS components and configurations, ensuring consistency and efficiency across large-scale production.

- **Lean Manufacturing Principles**: Implementing lean manufacturing principles helps optimize production processes and minimize waste. Techniques such as Just-In-Time (JIT) production, where materials are delivered as needed rather than stockpiled, and Six Sigma methodologies, which focus on reducing defects and improving process efficiency, can be employed to enhance scalability. By continuously analyzing and improving manufacturing processes, companies can ensure that scaling up does not lead to inefficiencies or quality issues.

1.2. Supply Chain Management

Effective supply chain management is crucial for scaling up BIS production. This involves coordinating the procurement of raw materials, components, and subassemblies to ensure that production schedules are met without delays.

- **Supplier Relationships**: Establishing strong relationships with reliable suppliers is essential for maintaining a steady flow of high-quality materials. Long-term partnerships with suppliers can lead to better terms, priority access to resources, and collaborative problem-solving. Diversifying the supply base also reduces the risk of disruptions due to supplier issues.

- **Inventory Management**: Implementing advanced inventory management systems helps manage the flow of materials and components efficiently. Technologies such as inventory tracking software and automated replenishment systems ensure that the right quantities of materials are available when needed, reducing the risk of production stoppages or excess inventory.

- **Logistics and Distribution**: Planning for logistics and distribution is integral to scaling production. Efficient logistics strategies, including optimized transportation routes and warehousing solutions, ensure timely delivery of BIS products to various markets. Advanced logistics technologies, such as real-time tracking and predictive analytics, can enhance supply chain visibility and responsiveness.

2. Strategies for Mass Production and Distribution

2.1. Production Techniques

Mass production of BIS involves scaling up manufacturing processes to produce large quantities efficiently. Several production techniques can be employed to achieve this:

- **Continuous Flow Production**: This technique involves setting up a production line where components move continuously through various stages of assembly. Continuous flow production minimizes downtime and maximizes throughput, making it suitable for high-volume manufacturing of BIS systems. Automated systems and standardized processes contribute to the efficiency and consistency of this approach.

- **Batch Production**: In batch production, BIS systems are produced in groups or batches rather than continuously. This method allows for flexibility in production, as different batches can incorporate variations or updates. Batch production is useful for managing production volumes and adjusting to changing market demands while maintaining quality control.

- **Cellular Manufacturing**: Cellular manufacturing organizes production into cells or workstations, each responsible for specific tasks or components. This approach enhances efficiency by reducing the time and distance between operations. Cellular manufacturing supports flexible production and quick adaptation to changes in BIS design or production requirements.

2.2. Quality Assurance and Control

Maintaining high-quality standards during mass production is essential to ensure the reliability and performance of BIS systems. Implementing robust quality assurance and control measures includes:

- **Standard Operating Procedures (SOPs)**: Developing and adhering to SOPs ensures consistency in production processes. SOPs outline the steps and criteria for each phase of manufacturing, including assembly, testing, and inspection. Following SOPs helps reduce variability and maintain product quality.

- **Inspection and Testing**: Rigorous inspection and testing protocols are crucial for identifying defects and ensuring that BIS systems meet performance specifications. Techniques such as automated testing, in-process inspections, and final product evaluations help detect and address issues before products reach the market.

- **Feedback Loops**: Establishing feedback loops between production teams and quality control departments facilitates continuous improvement. Collecting and analyzing data on defects, performance issues, and customer feedback helps identify areas for enhancement and refine manufacturing processes.

2.3. Distribution and Market Penetration

Effective distribution strategies are essential for reaching target markets and ensuring that BIS products are available to customers.

- **Distribution Channels**: Identifying and establishing distribution channels, including direct sales, partnerships with distributors, and online platforms, is crucial for market penetration. Tailoring distribution strategies to different regions and customer segments ensures that BIS systems are accessible to diverse markets.

- **Market Analysis**: Conducting market analysis helps understand demand patterns, customer preferences, and competitive dynamics. This information informs distribution strategies, pricing, and marketing efforts. Adapting distribution approaches based on market insights ensures that BIS products align with customer needs and expectations.

- **Global Expansion**: For widespread market reach, considering global expansion strategies is important. This includes navigating international regulations, adapting products to meet local standards, and establishing partnerships with international distributors. Global expansion requires careful planning and investment in infrastructure to support international operations.

2.4. Scalability Challenges and Solutions

Scaling up BIS production presents several challenges that need to be addressed proactively.

- **Resource Allocation**: Managing resources effectively is crucial for scaling production. This involves balancing the allocation of labor, materials, and equipment to meet increased demand. Strategic planning and resource management ensure that production capacity aligns with market needs.

- **Technology Integration**: Integrating new technologies and automation into existing production systems can be complex. Ensuring compatibility and minimizing disruptions during the integration process is essential. Investing in scalable technologies and conducting thorough testing can mitigate potential issues.

- **Workforce Training**: As production scales up, training the workforce to operate advanced machinery and adhere to quality standards is essential. Developing comprehensive training programs and providing ongoing support helps ensure that employees are equipped to handle increased production demands.

3. Conclusion

Scalability and mass production of Bullet Inversion Systems (BIS) involve a multifaceted approach encompassing modular design, automation, lean manufacturing, and effective supply chain management. Employing strategies such as continuous flow production, batch production, and cellular manufacturing supports efficient mass production. Ensuring high quality through robust assurance measures and addressing distribution challenges are crucial for successful market penetration. By addressing scalability challenges and leveraging innovative technologies, manufacturers can achieve efficient production and distribution of BIS systems, meeting the demands of diverse markets while maintaining product excellence.

Continuous Innovation and Research: Fostering a Culture of Progress

1. Encouraging a Culture of Continuous Innovation

1.1. Creating an Innovation-Driven Environment

Continuous innovation is a fundamental driver of progress in any field, particularly in technology-driven sectors like the development of Bullet Inversion Systems (BIS). To foster a culture of continuous innovation, organizations must create an environment that supports and encourages novel ideas and approaches.

- **Leadership and Vision**: Leaders play a crucial role in shaping an innovation-driven culture. By setting a clear vision that emphasizes the importance of innovation and progress, leadership can inspire and motivate teams to pursue creative solutions. Encouraging risk-taking and celebrating successes and failures alike fosters an atmosphere where innovation is valued and pursued actively.

- **Collaborative Workspaces**: Designing physical and virtual workspaces that promote collaboration and creativity can significantly enhance innovation. Open-plan offices, collaborative meeting rooms, and digital platforms for idea sharing enable team members to brainstorm and develop new concepts. Encouraging interdisciplinary collaboration brings together diverse perspectives, leading to more holistic and innovative solutions.

- **Incentive Programs**: Implementing incentive programs that reward innovative contributions can drive motivation and engagement. Recognition, financial rewards, and career advancement opportunities for employees who contribute groundbreaking ideas encourage a proactive approach to problem-solving and creativity.

1.2. Investing in Research and Development

A strong commitment to research and development (R&D) is essential for continuous innovation. Organizations must allocate resources and support R&D efforts to stay ahead in technology and meet evolving market needs.

- **Funding and Resources**: Allocating substantial funding to R&D activities enables the exploration of new technologies, materials, and methodologies. Providing access to cutting-edge tools, laboratories, and expertise supports the development of innovative solutions. Financial investment in R&D projects and infrastructure is crucial for maintaining a competitive edge.

- **Partnerships and Collaborations**: Collaborating with universities, research institutions, and industry partners enhances R&D capabilities. These partnerships provide access to specialized knowledge, advanced technologies, and additional resources. Joint research initiatives and shared projects accelerate innovation and bring diverse insights to the development process.

- **Fostering Talent**: Attracting and retaining top talent in science, engineering, and technology fields is vital for driving innovation. Investing in professional development, continuous learning opportunities, and career growth ensures that team members are equipped with the latest skills and knowledge to contribute to innovative projects.

2. Keeping Abreast of Technological Advancements

2.1. Monitoring Emerging Technologies

To remain competitive, organizations must stay informed about emerging technologies and trends that could impact their industry.

- **Technology Scouting**: Regularly scanning the technological landscape helps identify new developments and potential disruptions. Technology scouting involves monitoring research publications, patents, and industry news to stay updated on advancements relevant to BIS technology. Engaging with technology experts and attending conferences also provides insights into cutting-edge innovations.

- **Competitive Analysis**: Analyzing competitors' technological advancements and product offerings helps benchmark and identify areas for improvement. Understanding competitors' strategies and innovations provides valuable context for making informed decisions about the direction of one's own technology development.

- **Trend Analysis**: Identifying and analyzing trends in technology and market demand enables organizations to anticipate changes and adapt accordingly. Trend analysis involves evaluating patterns in technology adoption, consumer preferences, and industry shifts to guide strategic planning and innovation efforts.

2.2. Implementing Advanced Technologies

Adopting and integrating advanced technologies into BIS development processes can drive continuous innovation and enhance product performance.

- **Artificial Intelligence (AI) and Machine Learning (ML)**: AI and ML technologies can be leveraged to analyze vast amounts of data, optimize design processes, and predict performance outcomes. AI-driven simulations and predictive analytics improve the accuracy and efficiency of research and development, leading to more informed decision-making and innovation.

- **Advanced Manufacturing Techniques**: Incorporating advanced manufacturing techniques, such as additive manufacturing (3D printing) and precision engineering, enables the creation of complex and customized components for BIS. These techniques support rapid prototyping and iterative design, allowing for faster innovation and adaptation to new requirements.

- **Data Analytics and IoT**: Utilizing data analytics and Internet of Things (IoT) technologies enhances the monitoring and optimization of BIS systems. IoT sensors can collect real-time data on system performance, while data analytics provides insights into operational patterns and areas for improvement. This data-driven approach supports ongoing innovation and refinement.

3. Embracing an Iterative Development Approach

3.1. Agile Methodologies

Adopting agile methodologies promotes iterative development and continuous improvement in technology projects.

- **Iterative Prototyping**: Iterative prototyping involves developing and testing successive versions of a product or system. Each iteration incorporates feedback and lessons learned from previous versions, leading to incremental improvements and refinements. This approach allows for rapid adjustments and innovations based on real-world testing and user feedback.

- **Cross-Functional Teams**: Agile development relies on cross-functional teams that bring together expertise from various disciplines. By fostering collaboration among team members with diverse skill sets, organizations can address complex challenges and generate innovative solutions more effectively.

- **Customer Feedback**: Integrating customer feedback into the development process ensures that innovations align with user needs and preferences. Regularly engaging with end-users and incorporating their input into design iterations improves the relevance and effectiveness of BIS technology.

3.2. Continuous Improvement

A commitment to continuous improvement drives ongoing innovation and enhances the overall effectiveness of BIS technology.

- **Performance Metrics**: Establishing performance metrics and key performance indicators (KPIs) allows organizations to measure the success of innovation efforts and identify areas for improvement. Regularly reviewing and analyzing these metrics informs strategic decisions and guides future development initiatives.

- **Learning from Failures**: Embracing a culture that views failures as learning opportunities rather than setbacks fosters resilience and innovation. Analyzing and understanding the causes of failures provides valuable insights for refining processes, improving designs, and avoiding similar issues in the future.

- **Knowledge Sharing**: Encouraging knowledge sharing within the organization and across industry networks enhances collective expertise and supports continuous innovation. Hosting internal seminars, workshops, and knowledge-sharing sessions promotes the dissemination of insights and best practices.

4. Conclusion

Continuous innovation and research are crucial for advancing technologies like Bullet Inversion Systems (BIS) and maintaining a competitive edge. By fostering a culture of innovation, investing in R&D, staying informed about technological advancements, and embracing iterative development approaches, organizations can drive progress and achieve breakthrough solutions. A commitment to continuous improvement, coupled with a focus on emerging technologies and collaborative efforts, ensures that innovation remains a driving force in technology development.

Funding Models and Investment: Navigating Financial Strategies for BIS Development

1. Exploring Different Funding Models for BIS Development

Funding is a critical component in the development of Bullet Inversion Systems (BIS), which requires substantial resources for research, development, testing, and commercialization. To effectively finance BIS projects, various funding models can be employed, each offering distinct advantages and challenges.

1.1. Public Funding and Grants

Government Grants: Public funding through government grants is a traditional and vital source of financial support for technological advancements. Government agencies often provide grants to promote innovation, support scientific research, and address national security concerns. These grants typically do not require equity or repayment but come with stringent reporting and compliance requirements.

- **Research and Development Grants:** Specific grants are available for R&D projects, especially those aimed at advancing defense technologies or enhancing public safety. Applications for these grants usually involve detailed project proposals, including objectives, methodologies, and expected outcomes.
- **Defense and Security Funding:** BIS, being a defense-related technology, may qualify for funding from defense departments or security agencies. These agencies prioritize funding for technologies that enhance national security and improve military capabilities.

Public-Private Partnerships (PPPs): PPPs leverage resources from both the public and private sectors to fund technological innovations. In these partnerships, government entities collaborate with private companies to share costs, risks, and rewards associated with the development of new technologies.

- **Joint Ventures:** Public and private organizations may form joint ventures to develop BIS technology. This approach allows for pooling of resources, expertise, and funding, facilitating more rapid development and commercialization.

- **Innovation Hubs:** Governments may establish innovation hubs or incubators that provide funding, mentorship, and resources to promising technologies. These hubs often focus on advancing technologies with high potential for public benefit.

1.2. Private Funding and Investment

Venture Capital (VC): Venture capital is a significant source of private funding for early-stage technology companies. VC firms invest in high-risk, high-reward ventures with the potential for substantial returns.

- **Equity Investment:** In exchange for funding, venture capitalists typically receive equity in the company. This funding model aligns the interests of investors with the success of the technology, providing both financial support and strategic guidance.
- **Stage-Based Funding:** VC investments are often provided in stages, corresponding to different development phases of the technology. Initial funding may be used for prototyping and early testing, while subsequent rounds support scaling and commercialization.

Angel Investors: Angel investors are affluent individuals who provide capital to startups and emerging technologies. They often offer funding at an earlier stage than venture capitalists and may also contribute expertise and mentorship.

- **Personal Networks:** Angel investors frequently leverage personal networks and industry connections to support and guide the development of new technologies. Their involvement can provide valuable resources beyond financial investment.

Corporate Investment: Established companies in related industries may invest in BIS technology as part of their strategic initiatives. Corporate investors seek to integrate new technologies into their product portfolios or gain a competitive edge.

- **Strategic Partnerships:** Corporations may form strategic partnerships with BIS developers, providing funding in exchange for access to technology, exclusive rights, or joint development opportunities.
- **Acquisition:** In some cases, corporations may acquire emerging technologies or companies to accelerate their development and integration into existing product lines.

1.3. Crowdfunding and Alternative Funding Sources

Crowdfunding: Crowdfunding platforms allow individuals and groups to contribute small amounts of money to support technology development. This model democratizes funding and provides access to capital from a broad audience.

- **Rewards-Based Crowdfunding:** Contributors receive non-financial rewards or early access to the technology in return for their support. This model is often used to gauge market interest and build a customer base.

- **Equity Crowdfunding:** Equity crowdfunding allows contributors to invest in the technology in exchange for equity shares. This model attracts investors who are interested in the potential financial returns of the technology.

Grants and Competitions: Various organizations and foundations offer grants or organize competitions to support innovative technologies. These opportunities can provide non-dilutive funding and recognition for breakthrough developments.

2. Attracting Investment from Private and Public Sectors

2.1. Developing a Compelling Business Case

Market Potential: Demonstrating the market potential and commercial viability of BIS technology is essential for attracting investment. A thorough market analysis should highlight the demand for the technology, its competitive advantages, and its potential impact on the industry.

- **Value Proposition:** Clearly articulate the unique value proposition of BIS technology, including its benefits, differentiators, and potential applications. This helps investors understand why the technology is valuable and worth funding.
- **Business Model:** Present a well-defined business model that outlines revenue streams, pricing strategies, and financial projections. Investors seek clarity on how the technology will generate returns and achieve profitability.

2.2. Building a Strong Team

Expertise and Experience: Investors are more likely to support projects led by experienced and capable teams. Highlight the expertise and track record of the development team, including their skills, achievements, and relevant industry experience.

- **Leadership and Management:** Effective leadership and management capabilities are crucial for the successful execution of technology development projects. Showcase the team's ability to navigate challenges, manage resources, and drive progress.

2.3. Engaging with Investors

Networking and Outreach: Building relationships with potential investors requires proactive networking and outreach efforts. Attend industry conferences, participate in pitch events, and engage with investor networks to create opportunities for investment discussions.

- **Pitch Presentations:** Prepare compelling pitch presentations that effectively communicate the technology's value, development progress, and investment potential. Tailor presentations to address the specific interests and priorities of different investors.
- **Investor Relations:** Establishing strong investor relations involves maintaining open communication, providing regular updates, and addressing investor concerns. Building trust and demonstrating transparency are key to securing and retaining investment.

2.4. Navigating Regulatory and Compliance Issues

Compliance: Ensure that the BIS development process complies with relevant regulations and standards. Demonstrating a commitment to regulatory compliance reassures investors that the technology meets safety and performance requirements.

- **Risk Management:** Address potential risks and challenges associated with the technology, including regulatory hurdles, technical issues, and market uncertainties. Investors seek assurance that risks are identified and managed effectively.

2.5. Leveraging Funding Opportunities

Grants and Subsidies: Apply for relevant grants and subsidies to supplement private funding. This non-dilutive funding can reduce the financial burden and enhance the overall funding strategy.

- **Competitions and Awards:** Participate in technology competitions and awards programs to gain recognition and attract additional investment. Winning awards can validate the technology's potential and generate interest from investors.

Conclusion

Funding models and investment strategies play a crucial role in the development of Bullet Inversion Systems (BIS). By exploring diverse funding sources, including public grants, private investment, and alternative funding methods, organizations can secure the necessary resources to advance technology. Attracting investment requires a compelling business case, a strong team, effective engagement with investors, and adherence to regulatory standards. Through a strategic approach to funding and investment, BIS developers can achieve their goals and drive innovation in the field of personal protection technology.

Disaster Response and Humanitarian Applications: Harnessing BIS Technology for Crisis Management and Civilian Protection

1. Potential Use of BIS in Disaster Response Scenarios

Bullet Inversion Systems (BIS), while primarily conceived for defense and personal protection, offer intriguing potential for enhancing disaster response scenarios. In the context of large-scale emergencies, BIS technology could provide critical advantages in both immediate response and long-term recovery efforts.

1.1. Protection Against Debris and Projectiles

During disasters such as earthquakes, hurricanes, or explosions, the risk of flying debris and projectiles poses a significant threat to responders and civilians. BIS technology, by deflecting or neutralizing these hazards, could reduce injuries and fatalities.

- **Emergency Response Units:** First responders, including firefighters, paramedics, and search-and-rescue teams, often operate in highly hazardous environments. A BIS-equipped protective gear could shield these personnel from falling debris or shrapnel, enhancing their safety and operational effectiveness.
- **Civilian Safety:** BIS technology could be incorporated into protective shelters or emergency kits, providing civilians with an additional layer of safety during disaster events. This could be particularly valuable in scenarios where immediate evacuation or protection is critical.

1.2. Enhancing Rescue Operations

In rescue operations, the ability to provide safe passage through dangerous areas can significantly impact the effectiveness of the response. BIS technology could facilitate this by creating protective zones or barriers that allow rescuers to navigate hazardous environments more safely.

- **Securing Dangerous Areas:** BIS could be used to create temporary protective fields around unstable structures or hazardous zones, allowing rescue teams to operate with reduced risk. This application could be crucial in scenarios where traditional protective measures are insufficient or impractical.
- **Protective Barriers:** Deployable BIS technology could be used to set up protective barriers or shields that prevent debris from reaching rescue workers or survivors. This capability would be particularly beneficial in environments with high risks of falling objects or explosive debris.

1.3. Stabilizing Infrastructure

During and after a disaster, stabilizing compromised infrastructure is essential for ensuring safety and facilitating recovery. BIS technology could be employed to stabilize or reinforce structures temporarily, preventing further damage or collapse.

- **Structural Reinforcement:** BIS systems could be integrated into structural support frameworks to enhance their stability during aftershocks or ongoing hazards. This could help prevent additional collapses and protect both responders and affected populations.
- **Emergency Repairs:** BIS technology might assist in emergency repair efforts by providing temporary shielding or support for damaged infrastructure. This could be particularly useful in scenarios where conventional methods are hindered by ongoing hazards.

2. Humanitarian Applications for Protecting Civilians

The application of BIS technology extends beyond immediate disaster response to broader humanitarian efforts aimed at protecting civilians in various contexts. By leveraging BIS capabilities, several humanitarian applications can be explored to enhance safety and well-being.

2.1. Refugee Protection

In conflict zones or areas affected by natural disasters, refugees and displaced persons often face significant risks. BIS technology could provide protection for refugees by creating safe zones or enhancing the security of temporary shelters.

- **Safe Zones:** BIS systems could be used to establish secure perimeters around refugee camps or shelters, protecting inhabitants from external threats or hazards. This could improve safety and stability in areas with high risks of violence or environmental hazards.
- **Enhanced Shelter Security:** Incorporating BIS technology into the design of temporary shelters could provide an added layer of protection against potential threats, such as projectiles or explosive devices. This would enhance the overall security and resilience of humanitarian aid efforts.

2.2. Public Safety in High-Risk Areas

In regions prone to frequent disasters or conflict, BIS technology could be integrated into public safety initiatives to provide additional protection for civilians. This could involve deploying BIS systems in high-risk areas to mitigate threats and improve overall safety.

- **Disaster Preparedness:** BIS technology could be part of disaster preparedness programs, equipping communities with protective measures for potential emergencies. This would enable quicker and more effective responses to disasters, reducing the impact on affected populations.
- **Civil Defense Measures:** BIS systems could be employed in civil defense strategies to safeguard critical infrastructure and public spaces. By creating protective barriers or fields, BIS technology could help mitigate risks associated with natural disasters, terrorism, or other threats.

2.3. Post-Disaster Recovery

The recovery phase following a disaster involves rebuilding and restoring affected communities. BIS technology could support these efforts by providing protection during reconstruction and facilitating the stabilization of damaged areas.

- **Reconstruction Support:** During reconstruction, BIS technology could be used to provide temporary protection for construction workers and infrastructure. This would reduce risks associated with ongoing hazards and facilitate a safer rebuilding process.
- **Long-Term Protection:** In the long term, BIS systems could be integrated into new construction projects or infrastructure upgrades to enhance resilience against future disasters. This proactive approach would improve community safety and preparedness for potential emergencies.

3. Challenges and Considerations

While the potential applications of BIS technology in disaster response and humanitarian contexts are promising, several challenges must be addressed:

3.1. Technical Limitations

BIS technology must be adapted to effectively address the specific needs of disaster response and humanitarian applications. This involves refining the technology to handle diverse scenarios, such as varying types of debris or environmental conditions.

- **Adaptability:** Ensuring that BIS systems can operate effectively in different environments and situations is crucial for their success. This requires ongoing research and development to enhance adaptability and performance.
- **Integration:** BIS technology must be seamlessly integrated into existing disaster response and humanitarian frameworks. This involves coordination with other protective measures and systems to ensure comprehensive coverage and effectiveness.

3.2. Cost and Resource Allocation

Implementing BIS technology in disaster response and humanitarian efforts involves significant costs. Balancing the investment in technology with available resources and ensuring cost-effectiveness are essential considerations.

- **Funding:** Securing funding for the development and deployment of BIS technology requires strategic planning and collaboration with funding agencies, governments, and humanitarian organizations.
- **Resource Management:** Effective resource management is crucial for maximizing the impact of BIS technology. This includes optimizing deployment strategies and ensuring that resources are allocated where they are most needed.

Conclusion

The application of Bullet Inversion Systems (BIS) technology in disaster response and humanitarian contexts offers significant potential for enhancing safety and protection. By addressing challenges and leveraging the technology's capabilities, BIS can contribute to more effective disaster management and improved humanitarian efforts. Continued research, development, and collaboration are essential for realizing these benefits and ensuring that BIS technology meets the needs of diverse scenarios and populations.

Public Policy and Advocacy: Shaping the Future of Bullet Inversion Systems (BIS)

1. Engaging with Policymakers to Support BIS Adoption

1.1. Building Relationships with Policymakers

Engaging with policymakers is crucial for gaining support for Bullet Inversion Systems (BIS) and ensuring its integration into public safety strategies. Establishing strong relationships with government officials, legislators, and regulatory agencies can pave the way for favorable policies and funding opportunities.

- **Understanding Policy Makers' Priorities:** Policymakers often focus on issues that resonate with their constituents or align with broader governmental goals. It is important to align BIS technology with these priorities, demonstrating how it can address pressing concerns such as public safety, disaster preparedness, or national security.

- **Providing Evidence-Based Arguments:** To gain support, it is essential to present well-researched, evidence-based arguments that highlight the benefits of BIS technology. This includes providing data on its effectiveness, cost-efficiency, and potential impact on public safety. Detailed case studies, pilot project results, and comparative analyses with existing technologies can strengthen these arguments.

- **Engaging in Continuous Dialogue:** Maintaining an ongoing dialogue with policymakers ensures that they remain informed about the latest developments in BIS technology. Regular updates, briefings, and strategic meetings can help keep BIS on their radar and build a sense of urgency and importance around its adoption.

1.2. Navigating Regulatory and Legislative Processes

Understanding and navigating the regulatory and legislative processes are key to facilitating the adoption of BIS technology. This involves working within the framework of existing laws and regulations while advocating for changes that support BIS integration.

- **Compliance with Existing Regulations:** Ensuring that BIS technology complies with current regulations is essential for gaining approval and support. This includes adhering to safety standards, performance metrics, and any other relevant requirements.

- **Proposing Regulatory Adjustments:** If existing regulations pose barriers to BIS adoption, proposing adjustments or new regulations that facilitate its integration can be effective. This may involve drafting legislative proposals, collaborating with legal experts, and working with advocacy groups to push for policy changes.

- **Engaging in Policy Advocacy:** Partnering with advocacy organizations, think tanks, and industry groups can amplify efforts to influence policy. By leveraging these relationships, stakeholders can work together to advocate for BIS adoption and address any regulatory challenges.

2. Advocacy Strategies for Promoting BIS Technology

2.1. Public Awareness Campaigns

Raising public awareness about the benefits and applications of BIS technology can build support and drive adoption. Effective public awareness campaigns can educate various audiences, including consumers, businesses, and policymakers.

- **Educational Outreach:** Creating educational materials, such as brochures, infographics, and online content, can help inform the public about BIS technology. Workshops, seminars, and webinars can also provide opportunities for direct engagement and education.

- **Media Engagement:** Leveraging traditional and digital media platforms can help disseminate information about BIS technology to a broader audience. This includes press releases, news articles, social media campaigns, and interviews with industry experts.

- **Success Stories and Testimonials:** Highlighting success stories, case studies, and testimonials from users and experts can provide compelling evidence of BIS technology's effectiveness and benefits. These narratives can resonate with the public and build credibility for the technology.

2.2. Engaging with Industry Stakeholders

Collaboration with industry stakeholders, including technology companies, defense contractors, and research institutions, can support BIS advocacy efforts. These partnerships can provide valuable resources, expertise, and credibility.

- **Forming Alliances:** Establishing alliances with industry leaders and organizations can strengthen advocacy efforts and increase the visibility of BIS technology. Joint initiatives, collaborative projects, and shared resources can enhance the overall impact.

- **Leveraging Industry Expertise:** Engaging with industry experts can provide valuable insights and support for BIS technology. Expert endorsements, participation in industry events, and involvement in research collaborations can lend credibility and influence policy discussions.

- **Securing Endorsements:** Obtaining endorsements from respected organizations and influential figures can significantly impact advocacy efforts. These endorsements can help build trust and support for BIS technology among policymakers and the public.

2.3. Engaging with Advocacy Groups and Non-Profits

Advocacy groups and non-profit organizations focused on public safety, technology, and innovation can play a key role in promoting BIS technology. Collaborating with these organizations can enhance outreach and advocacy efforts.

- **Partnering with Advocacy Groups:** Working with advocacy groups that align with BIS technology's goals can provide additional support and resources. These partnerships can help amplify messages, organize events, and mobilize public support.

- **Leveraging Non-Profit Networks:** Non-profit organizations often have established networks and influence within their communities. Engaging with these networks can facilitate introductions to policymakers, support public awareness campaigns, and provide a platform for advocacy.

- **Aligning with Shared Goals:** Identifying and aligning with the shared goals of advocacy groups and non-profits can strengthen collaboration. By demonstrating how BIS technology supports their missions, stakeholders can create mutually beneficial partnerships.

3. Measuring and Demonstrating Impact

To effectively advocate for BIS technology, it is essential to measure and demonstrate its impact. Providing evidence of BIS technology's effectiveness and benefits can persuade policymakers and stakeholders of its value.

- **Data Collection and Analysis:** Collecting and analyzing data on BIS technology's performance, safety, and cost-effectiveness is crucial. This data can be used to create comprehensive reports and presentations that highlight the technology's impact and advantages.

- **Case Studies and Pilot Projects:** Documenting case studies and pilot projects that showcase successful implementations of BIS technology can provide compelling evidence of its benefits. These case studies can be used to illustrate real-world applications and outcomes.

- **Feedback and Improvement:** Gathering feedback from users, stakeholders, and experts can help refine BIS technology and improve its effectiveness. Demonstrating a commitment to continuous improvement and responsiveness to feedback can enhance credibility and support.

4. Conclusion

Public policy and advocacy are critical components in advancing Bullet Inversion Systems (BIS) technology and ensuring its successful adoption. Engaging with policymakers, leveraging advocacy strategies, and demonstrating impact are essential for gaining support and driving integration. By building relationships, navigating regulatory processes, and collaborating with industry stakeholders and advocacy groups, BIS proponents can effectively promote the technology and contribute to its successful implementation. As BIS technology evolves and matures, continued advocacy and strategic engagement will be crucial for realizing its potential and enhancing public safety and security.

Global Collaboration and Standards: Advancing Bullet Inversion Systems (BIS) through International Efforts

1. Encouraging Global Collaboration on BIS Research

1.1. The Necessity of Global Collaboration

Global collaboration in Bullet Inversion Systems (BIS) research is imperative due to the multidisciplinary nature of the technology, which spans physics, engineering, materials science, and more. The complexity and scale of BIS innovations necessitate a concerted effort from international researchers and institutions. Global collaboration enhances the collective expertise, resources, and technological capabilities available for BIS development, enabling advancements that individual entities might struggle to achieve alone.

- **Interdisciplinary Integration:** BIS research involves various scientific and engineering disciplines. Collaborating internationally allows researchers to integrate knowledge from different fields, leading to innovative approaches and solutions. For example, combining expertise in electromagnetism with advanced material science can yield breakthroughs in magnetic field generation and bullet deflection.

- **Resource Sharing:** Not all research institutions have access to the same level of resources. By collaborating globally, researchers can share facilities, data, and materials. This resource sharing accelerates development by providing access to high-performance laboratories, advanced simulation tools, and unique materials.

- **Addressing Global Challenges:** BIS technology has potential applications in global security and disaster response. Addressing such challenges requires a unified approach, leveraging diverse perspectives to tackle common issues. For example, global collaboration can help develop BIS systems that are effective in varied environmental conditions, from urban settings to extreme climates.

1.2. Strategies for Fostering Global Collaboration

To foster effective global collaboration on BIS research, several strategies can be employed:

- **International Research Networks:** Establishing international research networks can facilitate collaboration among experts from different countries. These networks can organize joint research projects, conferences, and workshops, fostering communication and cooperation.

- **Collaborative Platforms and Tools:** Utilizing digital collaboration platforms and tools can streamline international research efforts. Online databases, project management systems, and virtual collaboration environments enable researchers to work together more efficiently, regardless of geographical location.

- **Funding and Resource Sharing:** Securing funding from international grants and collaborative funding programs can support global research initiatives. Multi-national research grants can provide the necessary financial resources for joint projects and facilitate the sharing of resources among partners.

- **Joint Publications and Intellectual Property:** Encouraging joint publications and collaborative intellectual property agreements can promote knowledge sharing and recognition. This approach ensures that all contributing parties benefit from the advancements made and fosters a spirit of shared achievement.

2. Developing International Standards for BIS Technology

2.1. The Importance of International Standards

International standards play a crucial role in the development and deployment of BIS technology. They provide a framework for ensuring consistency, safety, and interoperability across different systems and applications. Developing international standards for BIS technology is essential for several reasons:

- **Consistency and Quality Assurance:** Standards ensure that BIS technologies meet consistent quality and performance criteria. This consistency is crucial for user confidence and safety, as well as for integrating BIS systems into existing infrastructures.

- **Facilitating Market Adoption:** Standardization can facilitate market adoption by providing clear and universally accepted criteria for BIS technology. This clarity helps manufacturers and users understand the requirements and capabilities of BIS systems, promoting broader acceptance.

- **Ensuring Safety and Reliability:** Standards define safety protocols and performance benchmarks, which are vital for minimizing risks and ensuring the reliability of BIS systems. By adhering to established standards, developers can enhance the safety and effectiveness of their technologies.

2.2. Processes for Developing International Standards

Developing international standards for BIS technology involves several key processes:

- **Formation of Standards Committees:** Establishing committees composed of experts from various fields and countries is essential for developing standards. These committees work collaboratively to define technical specifications, safety requirements, and performance criteria for BIS technology.

- **Comprehensive Reviews and Consultations:** Conducting thorough reviews of existing technologies, research findings, and best practices is critical for developing effective standards. Engaging with a broad range of stakeholders, including researchers, industry professionals, and regulatory bodies, ensures that the standards address practical concerns and emerging trends.

- **Collaboration with Standards Organizations:** Partnering with established standards organizations, such as the International Organization for Standardization (ISO) or the International Electrotechnical Commission (IEC), can provide guidance and support in developing and implementing standards. These organizations offer expertise and a structured process for creating and validating standards.

- **Iterative Development and Feedback:** Gathering feedback from industry experts, researchers, and end-users on draft standards is crucial for refining and improving them. Iterative revisions based on this feedback help ensure that the standards are relevant, practical, and reflective of the latest advancements.

2.3. Challenges in Standardization and Solutions

Several challenges can arise in the process of standardizing BIS technology:

- **Harmonizing Regulatory Requirements:** Different countries have varying regulatory requirements and safety standards. Harmonizing these regulations to create a unified set of international standards can be complex. Collaborative efforts among regulatory bodies and industry groups can help address these challenges and promote alignment.

- **Keeping Pace with Technological Advancements:** The rapid pace of technological advancements can make it challenging to keep standards up to date. Establishing mechanisms for regular review and updating of standards ensures they remain relevant and reflective of the latest developments.

- **Balancing Innovation with Standardization:** Ensuring that standards do not stifle innovation while providing clear guidelines can be difficult. Developing flexible standards that accommodate technological advancements while maintaining essential safety and performance criteria is key to addressing this challenge.

2.4. Benefits of Effective Standardization

Effective standardization of BIS technology offers several benefits:

- **Enhanced Safety and Reliability:** Standardized safety protocols and performance benchmarks ensure that BIS technology meets high safety and reliability standards, reducing risks for users and enhancing overall performance.

- **Increased Market Access:** International standards facilitate market access by providing clear criteria for technology evaluation and integration. This can lead to increased adoption and commercial success for BIS systems.

- **Improved Collaboration and Integration:** Standardization promotes collaboration by providing a common framework for technology development and evaluation. It also facilitates the integration of BIS technology into existing systems and applications, enhancing its effectiveness and usability.

3. Conclusion

Global collaboration and the development of international standards are crucial for advancing Bullet Inversion Systems (BIS) technology. By fostering international partnerships, leveraging collaborative platforms, and establishing effective standards, stakeholders can drive innovation, enhance safety, and facilitate widespread adoption of BIS technology. Addressing challenges and embracing opportunities in global collaboration and standardization will play a key role in shaping the future of BIS and its impact on public safety and security.

Closing Remarks:

The Bullet Inversion System (BIS) is an advanced technology designed to intercept and deflect projectiles using electromagnetic and plasma-based defenses. Its development draws from fundamental electromagnetism, including the Lorentz Force Law and Maxwell's Equations, to create effective magnetic fields that influence projectile trajectories. BIS technology also leverages plasma physics to form high-energy shields capable of absorbing or deflecting bullets.

Key areas of focus include the science behind magnetic fields, the properties of ferromagnetic materials, and challenges with non-magnetic projectiles. Power requirements for BIS are substantial, necessitating innovations in high-energy batteries and supercapacitors. Miniaturization and portability are crucial for practical applications, involving compact magnetic field generators and plasma systems.

Testing phases involve simulations, prototype development, and field testing to ensure BIS performance and safety. Improvements in energy efficiency, material durability, and future research into advanced materials are critical for enhancing BIS technology. Integration with protective gear and automated threat detection systems further extends its capabilities.

Funding models and investment strategies, along with regulatory compliance and ethical considerations, play significant roles in BIS development. Public policy, cultural attitudes, and global collaboration are essential for advancing the technology. Finally, scaling for mass production and continuous innovation are vital for realizing BIS's full potential and addressing future challenges.

This comprehensive approach underscores the BIS's potential to transform personal protection technology through ongoing research, development, and interdisciplinary collaboration.

www.ingramcontent.com/pod-product-compliance
Lightning Source LLC
Chambersburg PA
CBHW082234220526

45479CB00005B/1226